Mighty Mito

Power Up Your Mitochondria for Boundless Energy, Laser Sharp Mental Focus and a Powerful Vibrant Body

By Dr. Susanne Bennett

This book is intended to supplement, not replace,
the advice of a trained health professional, and
should not be considered a medical manual. It is not
a substitute for meeting with your doctor or to replace
or substitute any treatment that your doctor has prescribed.
If you know or suspect that you have a health problem,
you should consult a health professional. The author
and publisher disclaim any liability, loss or risk, personal
or otherwise, that is incurred as a consequence of
direct or indirect use and application
of any of the contents of this book.

Mention of specific products or companies
does not imply endorsement by the author or publisher,
nor does the mention of these products imply
that the companies endorse the book,
its author, or the publisher.

Internet addresses given were accurate
at the time this book went to press.

First edition: March 2016

ISBN: 978-0-9973735-2-3

Printed in the United States of America

1821 Wilshire Blvd., Suite 300
Santa Monica, CA 90403

www.mightymito.com
www.drsusanne.com

Dedication

This book is dedicated to those

who crave more energy

and ultimate vitality.

May you live a long and healthy life,

with much love and happiness,

free of pain and suffering.

Contents

Introduction

MY DISCOVERY OF THE POWER of the mitochondria began 5 years ago with the most severe injury I've ever experienced. It all happened as I was getting something out of my refrigerator that morning, and of course I was in such a rush trying to get to work, that's when an icepack fell out of the freezer, I went into a complete squat to pick it up then . . .

BAM!

I had stood straight up into an open upper refrigerator door. The pain was unbearable. My head and neck had crunched together like an accordion. At the moment of impact, I knew right away that I had injured my brain and neck severely because of the immediate shooting nerve pain and tingling radiating down both arms and legs . . . Being a doctor didn't help either because all I was thinking about was what can happen after such an acute injury. Not only was I concerned about the pain and swelling from the impact, but from the nerve pain and numbness down both my arms and legs. No, this was not just a big bump on my head . . . my symptoms pointed towards CNS issues. My central nervous system was definitely compromised, my brain and spinal cord, and as well as my PNS, peripheral nervous system. I had a TBI (traumatic brain injury) for sure!

The pain was so immense, I thought my life would never return to normal. Of course I rushed to work that morning with an icepack on top of my head, I had a full day of patients who relied on me so I couldn't let them down.

The shocking nerve pain down my legs subsided within a few days but the pain, numbness and tingling down both arms continued.

I started worrying: What if I can't work? How will I get around? What will I do with this pain and numbness for the next months, years? And then I seriously contemplated taking an early retirement from the work I love so much, which is being a chiropractic physician.

And here I was only 48 years old.

At the same time, I didn't actually realize the full extent of the traumatic brain injury (TBI) and just how deep the problems would run. It started affecting my hormones and my entire endocrine system very rapidly that within 4 weeks after the TBI, I lost my regular menstrual cycle (amenorrhea). My period stopped just like that and never came back.

The brain swelling from the acute injury was damaging enough that it started to choke my very delicate pituitary gland that was located deep in my brain, and within weeks, it started a cascade of symptoms, a condition known as *panhypopituitarism*. The pituitary is considered the master puppeteer of our body. It secretes eight hormones into our blood stream to control our major glands, and my TBI disrupted the function of three of them: my thyroid, adrenals and ovaries.

Due to the down regulation of my three major glands, along with my neurological symptoms, I was tired all the time, intolerant to a cold environment, and urinating excessively. Not only that, my hair was falling out, I had low blood pressure, I had imbalances in my blood sugar level and much more.

I had also developed *sarcopenia,* which is age-related muscle loss. While everyone loses some muscle as they get older, I was losing muscle at an alarming rate. I was prematurely aging within weeks of my brain and neck injury.

Emotionally, things couldn't have been worse. I had terrible anxiety, about the future, about my condition. I felt constantly on the edge. I couldn't get to sleep at night. I was a restless insomniac, and I started gaining weight and feeling sluggish and depressed. I've always been an extremely healthy and generally happy person because of the health protocols I've discovered and followed over the years.

This was a totally new and nightmarish reality for me.

I'm a big believer in both Western medicine and alternative therapies. So I started working with cranial sacral chiropractors, neurologists, Egoscue experts (postural retraining therapists), cold laser experts, acupuncturists, decompression specialists, and reflexologists. I did Korean Yoga, went to traditional healers, and neuromuscular reeducation specialists. I even bought different physical therapy machines to try to heal my neck and decrease my pain.

Now, a lot of these therapies were helpful, although some of them didn't work at all for my specific injury. That doesn't invalidate them. It just means that I needed something else, something *even more powerful* if I was going to heal myself this time around.

So in the next several months I read medical books and scientific studies, tried new therapies and supplements, and incessantly studied everything I could find about healing TBIs. I ended up learning a great deal about my body and about the root causes of disease and cellular damage. And not just about how the injury caused chronic inflammation in the brain, but more importantly, what was really going on at the cellular level. Molecular medicine was my new love!

And I learned what is perhaps the deepest form of healing possible, because it takes place at the very core of our cells, in the energy centers of our bodies: **the mitochondria.**

Why was this? Why were the mitochondria the key to the puzzles that neurologists, acupuncturists and chiropractors couldn't see?

The mitochondria are the place in our cells where our bodies produce energy. Let that sink in. Practically every molecule of oxygen, water, and food will need to go to the mitochondria so that we can laugh, play, cuddle, dance, and run. But we also need the mitochondria's energy for our bodies to breathe, digest, and pump blood, think, and feel.

And we need the energy from the mitochondria to **heal all pain and disease.**

Because of my TBI, I had severely damaged the mitochondria in my brain—one of the most important centers of these powerhouses of the body. So while many different therapies can help our bodies, only a holistic approach focused specifically on repairing damaged mitochondria and producing healthy ones could help me. Once I understood how to help my body rid itself of damaged mitochondria and create healthy new ones, I could heal the damage done from my traumatic injury.

Now I began to understand, because of my medical background—exactly what it would take at the level of nutrition, exercise, sleep, and also do more to provide my mitochondria with the nutrients and physical environment they needed to do their healing work.

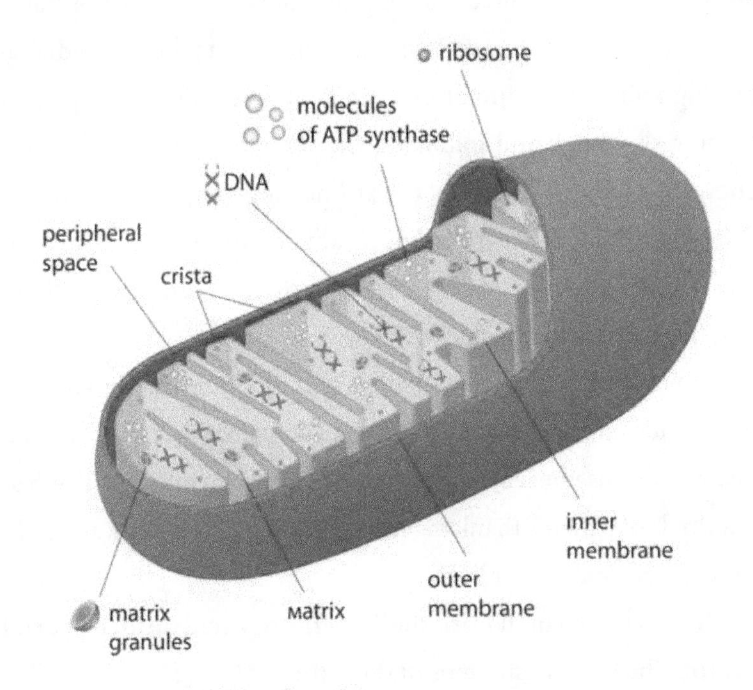

Mitochondrion

Once I began the very program that I'm going to outline in this book, my healing accelerated dramatically improved. My sarcopenia, which is the muscle loss I was experiencing—stopped in its tracks and I started to gain back the muscles I lost throughout my face and body. I was no longer aging rapidly and it felt like I was reversing my biological age! The anxiety, insomnia and stress went away. And today, 5 years later, I can say that I suffer *no* ill effects from the most painful injury in my entire life—an injury that could have left me incapacitated. Our mitochondria, those powerhouses of the cell—are *that* important. By healing them we can help reverse some of our very worst physical problems.

Only now is science realizing just how important the mitochondria are in *all* disease. Whether you're struggling with diabetes, high blood pressure or even general fatigue, I guarantee that reading this book and following the steps to restore your mitochondria will go a long way toward healing you from the inside. But our mitochondria are so important that healing them can also go a long way toward preventing other diseases such as Alzheimer's, heart disease and cancer.

What I'm going to share with you is exactly what I share with my friends and loved ones when they're suffering from pain and disease. I've used every strategy in this book myself, and I've had hundreds of patients thank me for sharing these strategies and radically transforming their lives.

In this book you'll learn the **3 P's for optimal mitochondrial health**. The three basic steps on how to:

- Prevent mitochondrial impairment
- Purge damaged mitochondria
- Protect and produce healthy, new mitochondria

This way, you'll have the most and healthiest mitochondria in your body possible, firing on all cylinders, making more energy for your body. As we'll see, the key to these steps is making certain lifestyle changes that go deep down to the biological level to improve your mitochondria's ability to process food and oxygen and convert them to *energy*.

These include:

- Minimizing exposure to mitochondria-damaging molecules (free radicals)
- Eating healthful foods rich in antioxidants (to fight free radicals)
- Eating a protein and fat-rich oxygenating pudding to supercharge your mitochondria
- Avoiding foods that harm mitochondrial functioning
- Taking the ideal supplements for mitochondrial health
- Using exercise to eliminate damaged mitochondria
- Building muscle for higher concentrations of mitochondria
- Improving your breathing hygiene and oxygenation
- Practicing stress reduction techniques
- Getting proper sleep for mitochondrial regeneration

This proven program will work for you, as it has worked for countless patients of mine. Not only will it help with the short-term issues we face, from inflammation, fatigue and insomnia, but it can go a long way toward preventing premature aging, Alzheimer's, cancer and more.

Yes, the mighty mitochondria are *that* powerful. And it's in your power to restore them to their very best!

PART I

The Mighty Mitochondria

CHAPTER 1

The Truth about Mitochondria

THE CYCLE OF LIFE BEGINS with us as young children filled with seemingly endless amounts of energy, jumping around, laughing, and playing. As we grow into teenagers and in our early 20s, our bodies become fully developed and we attain our prime levels of energy. Then, beginning in our 30s, or if we are lucky enough perhaps in our 40s, we start to lose energy and feel tired, and our bodies break down. Inevitably, if we haven't suffered from arthritis, inflammation, depression and other disease, we will slowly develop the diseases of advanced age: diabetes, heart disease, Alzheimer's and cancer.

Or so the story goes.

But I'm here to tell you that it doesn't have to be this way. I've had countless patients who've felt lethargic and are in chronic pain, suffering from conditions that they felt were untreatable. Like terrible allergies or diabetes, who have had dramatic healing results by applying one simple principle that I'm sharing with you in this book.

To heal your body, you must restore your mitochondrial health.

Mitochondria Are The Key To Energy

Mitochondria are the energy centers of our cells, as they provide almost all the "fuel" your body uses to breathe, eat, run, laugh, fight disease, and age vibrantly and happily. Just like a light bulb gets energy from a plug in a socket, our bodies in a similar way, get energy from our mitochondria. However, instead of electricity coming through a wall, our bodies run on food, water and oxygen. And mitochondria are the place where food and oxygen are converted into energy.

Therefore, we need to do everything in our power to prevent toxins from damaging our mitochondria, get them the best nutrients and optimum levels of oxygen possible, and finally learn how to remove damaged ones.

More and more research is showing that some level of mitochondrial dysfunction lies at the root of nearly *every* disease. At the most simple level, when your mitochondria go bad, you produce less energy, so you have less energy to fight disease. Unhealthy mitochondria equals an unhealthy you (fact).

But here's the good news: we all have the power to restore our mitochondria. And restoring them to their prime functioning, thereby giving you the most energy possible—is the whole purpose of this book.

Now, the story I painted at the beginning of this chapter isn't a myth entirely. There's naturally a lot of wear and tear that comes with living to seventy or eighty that is inevitable. In fact, Kirk Stokel writes, citing a study by Linnae, Kovalenko and Gingold that the "muscle tissue of a 90-year-old man contained 95 percent *damaged* mitochondria compared to almost *no* damage in that of a 5-year-old."[1]

Just think about that! This 90-year-old man had ninety-five times the damaged mitochondria as the average 5-year-old! No wonder our kindergartens are so filled with life and vibrancy. They're mitochondrial warehouses in action. On the other hand, the 90-year-old man's energy centers on those mighty mitochondria, which were only at 5 percent healthy functioning rate.

But have you ever met that 70-year-old who can still complete marathons? That 80-year-old who is out laughing at the coffee shop, smiling, filled with a life and happiness that many of us haven't felt since our 20s? A 90-year-old man or woman who loved nothing more than hiking or taking a long walk every morning?

Guess what their secret is? **Better mitochondrial health throughout their lifetime.**

In other words, having 95 percent mitochondrial damage at a later age is *not* inevitable. We can take steps every day to improve our mitochondrial health, which will in turn help us in both the short-term and the long-term. Aging, as many scientists are beginning to see, is nothing but this very damage we do to our mitochondria and cells. It's mitochondria damage, not simply a number of years, but that which determines how much energy you have, since that's where energy is produced! If you think about it, it makes total sense.

With that said, genetics do play a role in mitochondrial health. One thing we should note is that this book doesn't cover life-threatening conditions, such as Leigh Syndrome, which usually appears in the first year of life and involves severely damaged mitochondrial DNA, and other neurological disorders. However, there are many genetic variants that normal people have with their regular (non-mitochondrial) DNA. These can interfere with what's known as the methylation pathway, which without getting too technical, governs a wide range of biochemical processes and contributes to our mood and our ability to detoxify our bodies, and deal with disease and inflammation. When this pathway is impeded, it can contribute to mitochondrial dysfunction, between 30 to 70 percent of the time. Methylation issues can lead to problems in fat metabolism, the metabolism of folate (a crucial B vitamin), and carnitine metabolism, a compound that plays an essential role in energy production by processing fat.

The Myth That Genes Are Destiny

Did you know your energy depends on your mother's energy? Human beings get most of the mitochondria genetics (DNA) from our mother's side! So while genetics play a role in our mitochondrial health, the truth is that optimal physical and emotional health isn't something that some people are "blessed with" and other people can never get. We are all born on a gradient. Some of us with better DNA for these methylation pathways and for the power of our mitochondria.

There was a popular myth, beginning in the '90s, that we were nothing more than a sort of pre-programmed genetic code that played out our destiny in real life. If we had the "gene" for alcoholism, we became alcoholic. If we had the "gene" for depression, we became clinically depressed. This biological reductionism, besides the fact that no scientist ever really supported it in its absolute form, it did quite a lot of harm to people. Cause many people stopped believing they could make real changes or that they could just blame bad health or emotional difficulties on their parents, or their genes.

But the truth is *your genes are not your destiny!*

More recent medical research has proven that just the opposite is true. While DNA gives us a template, we can actually shape how it is expressed over the course of our lives. Two new exciting fields have arisen with cutting-edge research into how our lifestyle choices can make a difference at the cellular and even sub-cellular level (at the level of the mitochondria).

The first is *nutrigenomics,* which examines how nutrients interact with DNA and determine the expression or prevention of various diseases. Second is the field of *epigenetics,* which is the study of how environmental factors contribute to certain genes being turned on or off. As Laura Beil writes, when the National Institute of Health earmarked $190 million for epigenetic research in 2008, "Government officials noted that epigenetics has the potential to explain mechanisms of aging, human development, and the origins of heart disease, mental illness, and many other conditions" and goes on to say that some researchers think that epigenetics may "turn out to have a greater role in disease than genetics."[2] In other words, no matter what gene you are born with, whether for mental illness or heart disease, you can make changes to reduce the likelihood that it will be expressed.

If you change your environment and lifestyle, you change your genetic "destiny." I wrote all about that in my bestselling book, *The 7-Day Allergy Makeover* (for more information, check the Resources section in the back of the book). Clean up seven areas of your life in 7 days, and you can change your life and health destiny!

Knowingly or not, every day we all make choices in our life that shape just how healthy and happy we will be. Your lifestyle choices do, in fact, control your DNA and that's a good thing. It means making healthy choices creates healthy bodies.

So how do we know what is really a healthy choice? How do we know we aren't just falling into another marketing trap?

Health fads come and go. Low-fat used to be the thing. Now we're discovering more than ever the importance of healthy fats and the ratio of Omega 3s to Omega 6s for brain and cellular function. We used to think that drinking milk was a *requirement* in our food pyramids, and now we see health stores filled with almond and coconut milk, and plenty of healthy people I know forgo these altogether! Some people think they need to jog 2 hours to get a "real workout." Others pack themselves with proteins and powders to build muscle, when there are plenty of exceptionally strong people who have *never* had a protein shake in their lives.

So how can we tell what's a fad and what contributes to real, long-lasting health? What if we switched our *perspective entirely*? What if instead of thinking about treating disease after the fact, we started to think about our body's *innate healing capacities,* our inborn power to create energy and vibrancy?

Then the answer we'd get would be this: **let's create the most powerful mitochondria we can!**

Let's go to the cellular, subcellular and *genetic* level to give our bodies the nutrients, environment and chemical resources they need. That way our bodies can produce the most energy to fight everyday diseases and conditions, as well as the long-term issues we might face.

Better mitochondria mean more energy for *everything*.

That's why the health protocols in this book are based on something universal: the actual biology of the mitochondria themselves. What foods are better for mitochondrial health? I'll let you know and tell you exactly *why* these super foods are so good. What type of exercise is best for the mitochondria? I'll give you some workout tips and explain why certain types of exercise can actually hurt your mitochondria. What chemicals and compounds hurt the mitochondria and how you avoid them? I'll make that clear, so you can reboot your body for optimal health.

Every tip is based on preventing mitochondrial impairment; purging damaged mitochondria and helping your body protect and produce healthy mitochondria.

By reading this book and applying its protocols, you will literally be *rebuilding your body* and creating a healthy new body. Your mitochondria are the innate healers and centers of energy you have within you. By restoring them to their optimal state, your new body can come alive.

So before proceeding, let's briefly learn about these wonderful energy powerhouses.

CHAPTER 2

What Exactly Are Mitochondria?

NATURE HAS AN AMAZING WAY of working in dynamic patterns and cycles that reinforce and support each other. One of the most profound cycles that surround us each day is the flow of oxygen and carbon dioxide into and out of our bodies.

For millions of years, plants breathed in carbon dioxide from the environment and put out vast quantities of oxygen into the atmosphere. At the same time, sea-dwelling animals such as fish became adept at extracting this oxygen from water, through their gills. Over time, as land-dwelling creatures evolved, their respiratory systems changed so that they could extract this oxygen directly from the air. Nose, throat, lungs: these parts of our bodies are perfectly adapted to the intake of oxygen, the very chemical that plants breathe out so abundantly here on the surface of the earth. Without plants we could not survive, and these plants thrive on the carbon dioxide we exhale.

But until you picked up this book, have you ever thought about why we actually need all this oxygen?

The answer, of course, is to produce energy. Energy for walking, talking, laughing, and yes, even breathing! Along with the oxygen we breathe, the food we eat and the water we drink give us the vast majority of the chemicals we need in order to survive.

And the place where all that energy get produced is the *mitochondrion* (singular) or *mitochondria* (plural). Mitochondria are what are known as *organelles,* which are cellular structures with a specialized function. Other examples of organelles you may have heard of are the nucleus (the "brain" of the cell) and the endoplasmic reticulum (say that ten times fast). Many cells have hundreds of mitochondria whereas in vital areas, such as the heart, some cells can have over a thousand of them.

The origin of the mitochondria in our cells is very interesting. About 1.5 billion years ago, cells and mitochondria were two different things entirely. Mitochondria were high-energy protobacteria that actually invaded what today we know as the cell membrane. This cell membrane, which did have a nucleus, provided protection for the mitochondrion so it could perform its energetic functions in safety.

Your cells are almost like two creatures coming together to form a super creature. However, the mitochondria have become the powerhouses of this union, performing functions for the rest of the enclosed cell in which it found protection. In this symbiotic way, it became another organelle within a specialized and increasingly complicated cell. And those trillions of cells make you who you are today.

For this reason however, mitochondria have a different genome (or DNA sequence) than the rest of the cell and it resembles bacteria in a lot of ways. The downside is that this mDNA sequence leaves it vulnerable to attack and mutation by many of the chemicals we expose ourselves to every day. Because of this DNA difference, our mitochondria get damaged very easily by *free radicals* (discussed in depth below), which are noxious agents that come from things we ingest, breathe in and touch.

These free radicals produce what's known as *oxidative stress* or "rust." Our mitochondria (and other parts of the cell) are actually rusting when exposed to certain chemicals and compounds. And because mitochondria have different DNA, **oxidative damage to the mitochondria DNA (mDNA) actually occurs at five to ten times the rate of normal DNA (cDNA)!** It's that much easier to damage the mitochondria than the nucleus of the cell. Furthermore, mitochondria lack what's known as *histones,* the protective wrapping of our DNA, which would protect them against some of this oxidative damage. What happens is that once attacked, they actually create even more oxidative stress in a vicious cycle.

In addition, it's been recently discovered that not only does mitochondria produce energy in the form of ATP (discussed below), but they are also responsible for killing off cells when the body needs to. This process is known as *apoptosis,* is a crucial part of our normal cell functioning, since there are often cells we need to get rid of when they are damaged or unhealthy. However, increasingly, we are coming to discover how when something goes wrong in the mitochondria, they may turn this process of apoptosis against *healthy* cells in the body, leading to a whole number of autoimmune diseases. In simple terms, when mitochondria get damaged, they start mistakenly sending a signal to kill off *good* cells instead of unhealthy ones.

Inefficient parts of the mitochondria can actually be eliminated in the body, in what's known as *mitochondrial fission* and the healthy parts can combine with other mitochondria in what's known as *mitochondrial fusion*. Exercise, as we'll see, is a key step in getting rid of damaged parts of the mitochondria. Otherwise, damaged mitochondria will often reproduce more damage to the healthy cells, leading to disease. This epigenetic change of the DNA within the lifespan of an organism, means that we aren't producing the healthiest cells possible, instead we're producing diseased ones. This is why we want the mighty mitos (i.e. mitochondria) to be thriving and healthy as possible.

Because of this, it's a good thing to cull or get rid of our damaged mitochondria, leaving space for new healthy ones. This process is called *mitophagy* and occurs normally in our bodies. But as we age and as we damage our mitochondria, this process slows down, leaving more unhealthy mitochondria around. The net result is that our bodies aren't culling or "eating" the bad mitochondria, as they should be. In the next part of the book I'll explain exactly how we can go down to the cellular level to increase the rate at which we eliminate bad and damaged mitochondria and produce healthy new ones.

For now, let's be sure we understand exactly how mitochondria function to produce energy, and the crucial role of nutrients and oxygen in their work. This way, you will truly appreciate all the work they do in producing our bodies' needed energy and the importance of *properly oxygenating* your cells. They are an amazing part of our internal ecosystem, and I'm sure that once you understand exactly how vital they are for your overall health, you'll want to do everything you can to nourish and regenerate them.

How Is Energy Produced By the Mitochondria?

The main function of mitochondria is *energy production*. And this energy gets used for almost everything we do, which is why when your mitochondria are damaged, you have less energy to combat with disease, allergens, inflammation and stress.

Mitochondria produce a chemical energy in the form of a molecule called *ATP*. ATP stands for *adenosine triphosphate* and is simply a particular combination of hydrogen, phosphorous, nitrogen and oxygen that is particularly good at transferring its energy to cells for important functions like breathing, moving, digestion and cellular reproduction. ATP is like a fuel for your body; consider it as the currency, the cells use to work and survive.

ATP production takes place through what's known as the *Krebs Cycle* (or *citric acid cycle*). Without getting too complicated, the Krebs cycle uses the oxygen we breathe to break apart the carbohydrates, fats, and proteins from our food in order to produce ATP. The first step is the breakdown of glucose into pyruvic acid and from there, pyruvic acid gets oxidized, releasing two important chemicals. The first is CO2, or carbon dioxide, which is a byproduct of energy generation that we breathe out for the plants to breathe in. Another byproduct that's released in this process in H20, or water, helping to hydrate our bodies through the very food we eat.

Normally, this cycle flows perfectly, as the oxygen we breathe in helps to break down the foods we eat in order to produce ATP. The more mitochondria we have functioning at their peak, the more oxygen we can use from our environment. That means that our metabolism is increased and we can cycle through all the key functions of life more quickly. Faster removal of waste products, Faster removal of dead tissues, Faster flushing out of lactic acid, Everything "bad" gets eliminated from your system when your metabolism flows quickly and you start to feel less pain, less swelling, less pressure and less tension. Best of all, your body sends a message to your brain that says "I am healing" and through the mind-body connection, your healing can increase.

But when we expose ourselves to hazardous foods, polluted water, UV rays, drugs or toxic chemicals, or when we don't take care of our body's physical needs, this basic cycle of life can get thrown off. Instead of providing the oxygen and nutrients that our mitochondria need, we starve them. The body uses oxygen to fight off toxins instead of to produce energy for other key bodily functions. And a chemical chain reaction starts, and that causes disruptive molecules to take over. We have less and less available oxygen, so we produce less and less energy, in the form of ATP. This can lead to all sorts of physical symptoms, including fatigue, brain fog, moodiness, headaches, stiffness, joint pain and a weakened immune system (and more illness).

Over the long run, if these "mighty mitos" are injured or weakened, you start to experience disease. Exposing your body to harmful chemicals that can produce mutations in the mDNA. The body isn't taking care of itself because the powerhouse of the cell isn't able to be powerful. Oxygen isn't being used for vibrant energy, it's used to cope. If we aren't properly *oxygenated,* we aren't properly living.

Why Exactly are Mitochondria So Important to Our Health?

Mitochondria are an important part of lots of cells in your body. We have trillions of mitochondria throughout our body, and as we know they're where almost all of the oxygen we breathe gets put to use: **90 percent** in fact. We all know that without oxygen, we would die within a few minutes. But without mitochondria, our bodies literally couldn't process this oxygen. As we saw above, in the Krebs Cycle, mitochondria also use the chemicals in food to produce ATP.

So what do you *feel* if your mitochondria aren't working well? Your metabolism slows down. That means you process the metabolic waste products less efficiently (so it builds and clogs up in the tissues and systems), more of your total percentage of ATP production is required to take care of basic bodily functions (instead of going toward vibrant health), and you have less ATP to be your best self!

At the most basic level, when your mitochondrial health is impaired, you may start to experience symptoms such as:

- Fatigue and physical weakness
- Pain and soreness
- Memory loss and lack of motivation
- Brain fog and depression
- Mood changes and feeling overwhelmed
- Headaches and migraines

- Stiffness, tight or cramping of the muscles
- Prolonged healing and recovery period

In my nearly 27 years of my holistic clinical practice, I've seen thousands of patients who just couldn't understand why they had persistent low energy, pain, and a feeling that they just weren't operating at their optimal state. Brain fog and headaches shouldn't just be part of the daily routine of life. So many people tell me that they think it's just a natural part of getting older. Nothing else could be further from the truth! You don't have to suffer just because you're getting older. Our bodies are naturally happy, vibrant, and energetic, if our mitochondria are too.

If you're not feeling that way, then no matter how old you are, it's a sign that something isn't going right in the body. After I examine my patients' symptoms and got an understanding of their lifestyle choices, I often see a repeated pattern: every day my patients, without knowing it, are doing damage at the deepest cellular level and weakening their mitochondria. And if you've suffered from any of these symptoms, you'll know just how uncomfortable it can be. The good news is that no matter how much damage you've done, you can restore your mitochondria to optimal health and produce the energy your body needs for all of its vital processes.

As you might guess, mitochondria are so important that it's not just daily discomforts that crop up when something goes wrong. Our mitochondria can run into all sorts of other problems as well, ranging from digestive problems to inflammation to allergies because they are linked to the performance of our entire system. Autoimmune diseases are also signs of a dysfunction at this deeper level. Remember, mitochondria also have a function in deciding which cells should be eliminated. When things go wrong with the mitochondria, they can send messages to start attacking our own bodies!

Finally, mitochondrial health also governs our bodies' ability to age in a healthy manner and live vibrantly the older we get. Mitochondrial dysfunction is also a contributing factor of the chronic diseases many people struggle with as they get older, such as type-2 diabetes, heart disease, high blood pressure, cancer, Alzheimer's, arthritis and other autoimmune diseases. As Garth Nicolson writes, "Loss of function in mitochondria, the key organelle responsible for cellular energy production, can result in the excess fatigue and other symptoms that are common complaints in almost every chronic disease."[3] His research has also been on the delicate nature of the mitochondrial membrane, and how lipid nutritional therapy may be the key to optimal mitochondrial health, to prevent and heal the damage caused by free radicals. We will talk about mitochondrial membrane lipid therapy later in detail.

Preventing Mitochondrial Damage at the Source: Free Radicals on the Loose

The main disruptors of our mitochondrial health are *free radicals*. You've probably heard of them when watching a health-related TV show or maybe even on the back of your tea box. And for good reason. Free radicals are highly reactive atoms that want to bond to the cells in our body. If you remember high school chemistry, you'll remember that some molecules are highly stable since they are sharing electrons at the outer "ring" or valence shells.

However, free radicals are specific molecules that are unstable because they are *not* bound and have a free electron at this outer valence shell. They are "free" in the sense that they are reactive and especially eager to bond with chemicals in our body. These renegade electrons will try to latch on to the nearest molecule they can find, but sometimes this is extremely harmful, especially when it's our mitochondrial DNA!

Sometimes, free radicals can be very helpful, for instance, when used to "attack" viruses and bacteria. We don't want to, nor could we get rid of them entirely.

In addition, mitochondria also generate their own free radicals (reactive oxygen species, ROS) during the process of making energy, in another words, they can be harmful to themselves.

This double edge sword is called the "oxygen paradox." While oxygen is essential for human life, too much of the free radical metabolite are toxic to our body and mighty mitos. Free radicals are both beneficial and harmful; it all depends on the level of free radicals and your ability to reduce the damage.

But when too many free radicals are produced, they start bonding in places that are detrimental to our health. That bonding process whereby free radicals join with our cells, can actually damage them quite a bit, even killing them. Our mDNA is especially vulnerable to the effect of these free radicals.

The process whereby free radicals bond to molecules in our body is what's known as *oxidative stress.* As mentioned before, it's like your cells begin to develop "rust," just like nails on an old fence that are exposed to the elements. Furthermore, free radicals often easily bond with oxygen, leaving less oxygen available cellular processes, like the production of ATP! The more free radicals that bond to oxygen, the less oxygen there is to make energy.

That's why we'll come back time and time again to this important saying: *for optimal mitochondrial health, you need optimal oxygenation.*

Mitochondrial DNA (mDNA) is five to ten times more prone to free radical attack as cellular DNA (cDNA). The free radicals that target mitochondrial DNA (mDNA) are what's known as *reactive oxygen species* (ROS) and cause mutations in the mDNA. When this mutated mDNA replicates, it will produce more damaged mitochondria instead of healthy ones. These mutations only increase as we get older, especially past age sixty-five. That's why it's crucial to counteract our exposure to free radicals through our diet, exercise program, and even breathing habits.

Mutations can occur in many different organs, but when it comes to mitochondrial dysfunction, it seems that the muscles and the brain are particularly vulnerable. Muscles typically have the highest concentration of mitochondria in them. This is also why we want to build muscles through exercise, for more mitochondria, and therefore, more energy. For this reason, if you're not treating your body right, muscles can be the first to be attacked. Hence feelings of soreness, pain and slow healing that come when you have a deep dysfunction. Remember my rapid loss of muscle mass caused by my head injury? I was losing healthy mitochondria as well!

Additionally, the brain is another key area that can suffer from mitochondrial damage or mDNA mutation. Both muscles and the brain require a lot of energy to function. The brain in fact, requires around 20 percent of the energy we produce in our bodies. The brain also requires a lot of oxygen and because it is composed of mostly fats, is especially vulnerable to attacks by free radicals and ROS. So you can also see where feelings of brain fog and disorientation might arise the minute we aren't taking care of those mighty mitos! Think of a hangover as a barrage of these free radicals to the brain.

CHAPTER 3

The Long-Term Effects of Free Radicals and Mitochondrial Damage

THE MORE FREE RADICALS WE'RE exposed to, the higher chance that we're going to have some damages done to our mitochondrial integrity. Over a long period of time, this can produce what we call the "diseases of civilization." These are the type of diseases that might not have afflicted our ancestors who lived to only thirty or forty, but are the most common threats to our health now that we live into our 70s, 80s, and 90s. Not only are many of them fatal, but they reduce our day-to-day wellbeing. The good news is by keeping our bodies properly oxygenated and our mitochondria thriving, we can slow down or even reverse some of these processes.

The overall level of oxidative stress in the body and long-term damage to the mitochondria has been linked to many health disorders as discussed by Dr. Steve Pieczenik and Dr. John Neustadt in their 2007 article in *Experimental and Molecular Pathology* titled, "Mitochondrial Dysfunction and Molecular Pathways of

Disease" (for more information, check the Resources section in the back of the book.)

These conditions include:

- Parkinson's and ataxia
- Alzheimer's disease
- Diabetes
- TIAs, or transient ischemic attacks (loss of blood flow to brain)
- Coronary artery disease and cardiomyopathy
- High blood pressure
- Chronic fatigue syndrome
- Fibromyalgia (muscular pain syndrome)
- Inflammatory diseases
- Neuropathic pain
- Migraines and other headaches
- Schizophrenia and bipolar disease
- Epilepsy
- Stroke
- Retinitis pigmentosa
- Liver diseases including hepatitis C and biliary cirrhosis
- Increased rates of sarcopenia (age-related muscle loss)

Since one of the favorite targets of free radicals is the mitochondrion, DNA and membrane mutations may be the root of all diseases. As these damaged cells reproduce, they create damaged mDNA that can lead to these diseases. Free radicals start off a chain reaction, going from one cell to the next, causing oxidative stress, impacting more and more of our bodies' DNA.

Alzheimer's

Alzheimer's in particular has been increasingly linked to excessive carbohydrate and sugar consumption, such that some experts have even begun to call it "Type III Diabetes." Deposits of a sticky substance called "beta-amyloid plaque" in the brain tissue is a telltale sign of Alzheimer's impaired brain physiology and is possibly caused by damaged mDNA.

The more powerful our mitochondria, the more we can burn fat and not overload the cells with the demands of breaking down simple carbs.

Diabetes

Speaking of diabetes, we know excessive sugar and carbs can lead to irregular blood-sugar levels and obesity. One of the byproducts of breaking down sugar in the body is lactic acid, it's the stuff that makes our muscles sore and increases the acidity in our body. If your mitochondria aren't functioning well, it's hard to get rid of that lactic acid as a metabolic waste product and that leads to further damaged cells.

However, what you might not realize is that because our muscles are particularly packed with mitochondria, it is easier to shuttle sugar, lactic acid, and other waste products out of the system and therefore better regulate glucose levels in the body. That's why part of what we're doing when we exercise is simply building better mitochondria. We're packing our body with them, since they're highly concentrated in muscle. They're truly that powerful, even when it comes to diabetes.

One very important note I would like to make about diabetes, after treating hundreds of patients with sugar imbalance issues, I believe that it's completely preventable and even reversible. And it all has to do with the amount of muscle you have! The more muscle mass you have, the higher chance of lowering that blood sugar level postprandial (after a meal) to prevent hyperinsulinemia (high levels of insulin) which will ultimately prevent insulin insensitivity of the tissues. Of course that means you have to use the muscles after that meal. You can't just be sitting around watching TV or listening to music. That is why I ask all of my patients to walk after each meal, for around 15–20 minutes, to use up that readily available glucose. More muscle means more mitochondria. Ten percent of muscle mass is made up of mitochondria.

Weeks after my TBI, I would say I lost up to 20–25 percent of my muscles. Interestingly, after the injury, my blood tests confirmed that I was having sugar imbalances as well. My risk of diabetes was going up, indicated by my Hemoglobin A1c (HA1c) levels. HA1c is a marker for risk of diabetes, and it gauges how well your body is handling the blood sugar levels in the past 3 months.

I have always had HA1c at around 5.1 to 5.2 and soon after my accident, dealing with my endocrine issues and rapid muscle loss, it rose up to 5.6, which is the highest number within the normal range. If one of my patients had a HA1c level of 5.6, I actually tell them that they are definitely in the pre-diabetes range. I was absolutely shocked how quickly my body was breaking down in just weeks! I don't want to get ahead of myself here, so know that I will discuss exactly how I was able to reverse my pre-diabetes and get my HA1c levels back down in the coming chapters!

Cancer

Cancer is one of the major health risks in our society. Many scientists theorize that the damage done to mitochondria produce mutations in DNA, leading to the reproduction of cancerous cells. If you take a look on lined at the list of toxins that trigger high levels of free radicals, you'll notice how many of them is from smoking cigarettes and ingesting foods with pesticides to breathing in toxic polluted air, are carcinogenic in nature. Further, while damaged mitochondria "turn on" the reproduction of cancerous cells, stronger mitochondria may be able to "turn off" their replication. In other words, **the healthier our mitochondria, the more likely we are creating the healthiest next generation of cells possible**. In fact, some biologists theorize that *aging itself* is simply the long-term effect of damage caused by free radicals in our system.

But there's another connection between cancer and mitochondria. Dr. Patrick Quillin has noted Dr. Otto Warburg's interesting connection between the ways that cancer cells survive, with the anaerobic process of fermentation, compared to normal cells:

The 1931 German Nobel laureate in medicine Otto Warburg, M.D., first discovered that cancer cells have a fundamentally different energy metabolism compared to healthy cells. The crux of his Nobel thesis was that malignant tumors frequently exhibit an increase in anaerobic glycolysis, a process whereby glucose is used as a fuel by cancer cells with lactic acid as an anaerobic byproduct, compared to normal tissues.[4]

In other words, cancer cells tend to feed on sugar through fermentation. And fermentation is an anaerobic activity, meaning *it doesn't use oxygen*! So the more we encourage our bodies to burn foods and turn them into ATP through the mitochondria, with oxygen, the more we may be "starving" cancerous cells of the sugar they need to thrive. Fermentation is a laborious process for the body to undertake and produces far, far less ATP than the aerobic processes of the mitochondria and the Krebs Cycle.

Put simply, oxygenation is life!

High Blood Pressure and Heart Disease

When it comes to high blood pressure and heart disease, we all know that cholesterol is a big culprit. But not for the reason most people think. In fact, it's an important hormone that's used in the adrenals, testes and ovaries to make other hormones, such as testosterone, cortisol and estrogen. The location where cholesterol converts into these important mineral corticoids, glucocorticoids and steroid hormones, is actually *in the mitochondria*!

So if the cholesterol can't be shuttled into the mitochondria due to damage, low mitochondrial numbers or oxidative stress, what do you think will happen to your cholesterol levels? Yes, it will go up! Cholesterol is not truly the bad; it's a vital molecule needed hormone production and more, but what needs to be understood is that the delivery and utilization by the mitochondria Is essential for optimal health!

Sarcopenia

Sarcopenia begins to set in during your 30s when you start to lose muscle mass as part of aging. However not everyone loses muscle mass at the same rate. When it comes to sarcopenia, sedentary lifestyles, low levels of protein intake, obesity, and high levels of oxidative stress can all contribute to rapid muscle loss and premature aging. That's right, too much exposure to free radicals makes you age more quickly.

Remember that 90-year-old man we discussed at the beginning of the book who had 95 percent of his mitochondria damaged? That's a good sign that his muscles have also degenerated and the aging process has gone "faster" than it has to for some people. And since our muscles are particularly packed with mitochondria, increased rates of sarcopenia have a doubly bad effect. Not only do we lose muscle that could be used to get around more easily, but we also lose the mitochondria that could *produce* more energy for strength and endurance. That's why avoiding and minimizing free radicals and reducing oxidative stress is increasingly important the older you get.

CHAPTER 4

The 3 Ps of Optimal Mitochondrial Health

SO WHAT'S THE SECRET TO good mitochondrial health? It's actually quite simple. Remember how we said that mitochondria uses a huge quantity of your body's oxygen, 90 percent in fact? The key to optimal mitochondrial health is to improve your body's ability to intake and use oxygen effectively. More and better oxygen use results in stronger, more powerful mitochondria, no matter how old you are.

Put simply, *for optimal mitochondrial health, you need optimal oxygenation.*

Now, that doesn't mean you need to rush off to an oxygen bar every day, although you will see how breathing properly and breathing purified air is important for your intake of oxygen. But it does mean that you have come to learn the ways that our lifestyles, food, sleep, water, exercise and supplements can help the mitochondria process oxygen.

By getting proper oxygenation, you can train your body to operate at optimum mitochondrial health. As mentioned, it's a three part-process in which you:

- Prevent mitochondrial impairment
- Purge damaged mitochondria
- Protect and produce healthy, new mitochondria

And best of all, you don't have to be a geneticist to make it happen. Although the healing goes on at a deep cellular level, it's really about cultivating a new mindset and new habits. I'll show you how implementing simple changes to your diet, sleeping patterns, and exercise program will greatly increase the amount of oxygen your mitochondria have to utilize. You'll also learn all about the best supplements for functioning optimally at this deep cellular level.

In addition, you'll learn how to avoid the *free radicals* and dangerous toxins that compromise mitochondrial health. These chemicals are literally all around us, in the air we breathe, the sun we enjoy, the water we drink and the food we eat. Exposure to free radicals can increase the "rust" in our bodies, making proper energy production more difficult. It can even lead to some diseases that develop over the long term, such as Alzheimer's, heart disease and cancer.

But there's good news. By supercharging your mitochondria, you can slow down or even reverse the damage done to your cells that's at the root of so many of these debilitating diseases. I'll show you how certain key foods, what I call *mitochondrial super foods* will charge up your cells and help these powerhouses get churning. The key is that these foods are packed with *antioxidants,* which are molecules that "fight off" free radicals, giving them no place to bond, instead of allowing them to bond to your mitochondria. You'll also learn about some of the key supplements you can start taking *today* that will boost your mitochondrial health and improve your body's ability to make use of oxygen. And that means a healthier, more energetic, happier you.

When it comes to diet, you'll see why a healthy dose of vegetables, proteins, and fats are ideal for your mitochondria's energy-synthesizing abilities, whereas sugar, which break down outside of them, are really just a short-term fix for your energy needs and that they can actually hinder your ability to make more mighty mitos.

You'll also learn the recipe for my supercharged *oxygenating pudding*: a tasty mixture of fats and proteins that your body can metabolize quickly to trigger increased oxygen use in your cells. This is also great news since diseases have a difficult time thriving in an oxygen-rich environment.

Finally, I'll show you how to improve your lifestyle, through improving your breathing (and therefore oxygenation), sleep, and exercise. Which will help your mitochondria recharge themselves and give you the most energy possible.

PART II

Let the Healing Begin!

CHAPTER 5

Step One: Prevent Mitochondria Impairment

LET'S FIRST LOOK AT SOME of the environmental pollutants replete with free radicals that cause mitochondrial dysfunction. By avoiding them or minimizing your exposure to them, you can prevent the damage that free radicals do to your mighty mitos. Toxins in the air, chemicals in our water and even the light we need can all impair our mitochondrial functioning. You'll learn exactly how mitochondria become damaged and denatured and what environmental toxins to avoid in order to prevent further damage in your system.

However, for a more thorough look at how environmental toxins can harm our bodies, including our mitochondrial health, check out my book *The 7-Day Allergy Makeover* available for purchase on online at The7DayAllergyMakeoverBook.com and Amazon.com as well as in your local bookstore.

So how do we mistreat our mighty mitos through free radical exposure and what can we do to heal them? The first part of the answer is in *minimizing* our constant exposure to free radicals.

While we can never avoid, nor would we want to avoid free

radicals entirely, there are some environments and habits that greatly increase our exposure to these free radicals. Part of living in civilization is a higher exposure to free radicals. As we'll see, smog, UV rays and chemical fumes are all around us and can produce mitochondrial dysfunction. But there are also a number of lifestyle choices that can minimize our exposure to free radicals. You might not be surprised to learn that processed foods end up producing far more free radicals than whole, natural, organic foods.

When it comes to our environment, especially in the air we breathe, free radicals are everywhere. No one can live in a cave or breathe fresh mountain air all the time. But think of what a difference it does make after that first day of hiking or a stroll on the beach of a remote island. Only then do you realize how much clearer your thinking is and how much more energetic you feel. Part of that is naturally the beautiful environment, but part of that is breathing air that is devoid of all the toxic chemicals and pollutants we breathe every day.

So where do airborne free radicals come from? We get these free radicals toxins from:

- Polluted air and smog
- Chemical fumes
- Mold spores and mycotoxins
- Gaseous Toxins (such as methane and carbon monoxide gas)
- Smoking tobacco, electronic cigarettes and marijuana

- Volatile Organic Compounds, or VOCs (such as formaldehyde and toluene)
- Perfume, air fresheners, potpourri

While a certain amount of exposure to these airborne toxins is inevitable, others can be avoided. These pollutants can actually change the number of copies of mDNA that the body makes as well as the expression of mDNA itself.[5] That's right, breathing polluted air can make you a mutant! That's why breathing pure air is the key.

Keep Air Clean

As I've discussed in my book, *The 7-Day Allergy Makeover,* using a HEPA air purifier in your house can help you avoid some of the worst airborne toxins, since it can filter out over 99.9 percent of all harmful chemicals in the air. Be sure to get a HEPA filter that is properly-sized for your room. You should measure the square footage of the room *before* buying the purifier, since the one that's put in a large room will need to be powerful enough to be able to cycle through all the air. Otherwise, you won't be getting the room completely clean.

Also, avoid imitations that say things like "HEPA-like" or "HEPA-type." These don't actually filter out the same number of particles and leave you exposed to free radicals that can damage your mitochondria. It can be confusing when you see all these names at the store, so make sure it says "HEPA" and nothing else. You'll also want to make sure it includes a charcoal filter to absorb harmful gases and release the purest air possible. It's very easy, especially if you keep your windows open, for smog and other pollutants to come into your home.

Another natural way to cleanse the air in your home or office is with live plants. It adds more clean oxygen to your surroundings and at the same time removes pollutants from the air. The most effective plants for removing formaldehyde and other VOCs from your indoor living environment include the golden pothos, spider plant, and philodendron. Just be sure not to overwater. This can cause mold to grow (another toxin you should avoid if possible).

Avoid Carbon Monoxide While Driving

When it comes to chemical fumes and toxins, you may not even realize the incredible amount of daily exposure! A drive on the freeway once in a while is fine, but how many of us are exposed to harmful chemicals, including those in smog, every day of our lives? One easy trick is to hit the recirculate air button in your car, so that you will not be taking in air from the outside. It's also a good idea to shut down your car and wait a few seconds to get out if you're parking in a garage. If it's your own garage, leave the garage door open to release the carbon monoxide as long as you can before leaving the car, especially if there are rooms in the house above or next to it. That carbon monoxide and car exhaust is mitochondrial poison and also harmful for to the rest of your body!

Avoid Chemical Beauty Products

Additionally, there are numerous toxins in many of the health and beauty products we use that can seep into our skin and damage our mDNA. I believe that for this reason, it's best to use only all-natural lipstick, blush, hand soap, shampoo and more. For generations, our ancestors only used plants, minerals, and other naturally-occurring objects in order to cleanse, exfoliate, and moisturize. The industrial revolution produced a ton of new products, but also many that should be stayed clear of, because of how they can interact at the mitochondrial level. That way, you'll avoid chemicals such as phthalates, parabens, oxybenzone, diethanolamine (DEA) and heavy metals (mercury and lead, used in lipstick), some of which, besides causing oxidative stress, may actually be carcinogenic. Many of these chemicals are so harmful, and need to be avoided as they are even banned in the European Union, but not yet in the United States. Whether or not a ban is coming, you can get rid of them in your own bathroom. If it doesn't sound all natural, don't use it!

Don't Smoke or Vape

Smoking is obviously one of the worst things you can do for your cellular and mitochondrial health and stopping is a necessity to offset not just damage to your lungs, but other types of health problems we'll discuss below. If you have switched to electronic cigarettes (e-cigarettes), I don't believe they are any better! Unlike tobacco cigarettes, e-cigarettes generally are battery-operated and use a heating element to heat up a liquid from a refillable cartridge, releasing a chemical-filled aerosol infused with nicotine and various other chemicals.

Did you know there is no regulation of e-cigarettes from the government or FDA? There are up to 500 brands and 7,700 flavors on the market, all without any form of authority determining if the ingredients are safe to breathe for short or long-term health. It's too soon to tell what the long term effects will be, since the e-cigarettes haven't been around long enough to be tracked, but we do know some ugly truths.

According to the American Lung Association, the initial studies conducted by the FDA in 2009 found detectable levels of toxic cancer-causing chemicals in two leading brands and 18 various cartridges. They also found chemicals that were common found in antifreeze. Yikes!

In 2014, researchers found that the vapors from e-cigarettes with a higher voltage level had higher levels of formaldehyde, a well-known carcinogen.

Science has shown us for years that every moment you inhale from a regular cigarette or a vaporizer, you are sucking down harmful toxins that can be dangerous to your health or possibly your unborn fetus. We know this because on every box of cigarettes there is a warning label. Some of you reading this right now are thinking that "I'll be fine, I have no symptoms. That won't happen to me!" I hope that is absolutely true, and I wish you the best of health. But I also know some of the readers are still smoking and may be suffering from a chronic lung disease such as COPD (chronic obstructive pulmonary disease) or even lung cancer. I am asking you to choose a life. Choose every day to be your very best, full of energy and mental clarity. Make a commitment today to quit smoking for yourself, your family, your future generation and planet Earth!

Avoid Volatile Organic Compounds

Volatile organic compounds, or VOCs, are harmful chemicals that are commonly found in car exhaust, solvents, drycleaners, particle board, glues and conventional paints as well as that "new car smell." VOCs are in fact one of the main components of smog as these seep into the environment. Inhaling these toxins can trigger a headache and leave you feeling woozy, disoriented and exhausted, which are all symptoms of mitochondrial dysfunction. For all these reasons, you want to breathe the cleanest air.

CHAPTER 6

Free Radicals in What We Ingest

ON THE OTHER HAND, we are also exposed to a number of free radicals based on what we eat, drink, and absorb. These include:

- Alcohol, recreation and illegal drugs
- Processed foods
- Pesticides, herbicides and fungicides on food
- Antibiotics and prescription drugs
- Heavy metals, chemicals and radioactive byproducts in tap water
- BPA (Bisphenol A) and plastics

While most food ends up producing some quantity of free radicals, there are some things we ingest in food and water that are particularly harmful.

Alcohol

Alcohol is the anti-mitochondrial chemical above all others! There's almost nothing that destroys healthy mitochondria as quickly and lowers your body's ability to produce ATP. While most of us enjoy a glass of wine or a beer every once in a while, you should really take a look at your alcohol consumption if you're concerned about the long-term health risks, including diabetes, progressive sarcopenia, obesity, Alzheimer's, heart disease and more that are all linked to mitochondrial dysfunction. Alcohol may put you at higher risk for all of these issues.

Processed Food

Likewise, the food choices we make are of utmost importance. Keep in mind that mitochondria can get overloaded and damaged by processed, refined foods, especially those with high carbohydrate (sugar) contents. That means foods like white bread, soda, sugars, and sweeteners are real mitochondria killers.

Pesticides, Herbicides and Fungicides

Additionally, eating conventionally grown foods, those sprayed with pesticides, can damage your cells. Organic isn't just a fad. It's crucial to your long-term health and vitality. With that being said, if fresh organic foods are just not an option for you whether it doesn't fit into your budget or not easily accessible, then check out your local farmers market and buy from a farmer who doesn't use pesticides or fertilize with chemicals. Also, frozen organic fruits and vegetables are a great alternative for soups and shakes. We'll spend a lot of time looking at the very best choices for your nutrition, and my *Ultimate Wellness Food Checklist* (the complete list is available in the Appendices and a download link in the Resources section in the back of the book) will let you know everything you can eat and what to avoid.

Antibiotics and Prescription Drugs

Antibiotics, however invaluable for certain diseases, are often overprescribed, leading to an array of health complications. For one, they can ruin your inner gut flora, the helpful bacteria that regulate digestion and even your mood and lead to increased bacterial resistance. However, antibiotics also increase oxidative stress in mitochondrial DNA (mDNA). In animal studies with mice, they have also been found to increase the expression of genes related to anti-oxidation, meaning that these mice were trying to fight off the greatly increased rates of oxidation.[6] Some prescription drugs can also have damaging effects and you should discuss these with your prescribing physician.

The most well known prescription drug that destroys the number and physiology of mitochondria are statin drugs. In fact, a 2010 report from the National Center for Health Statistics published that in the United States alone, one out of four Americans over the age of forty-five are taking a statin.

Doctors prescribe statins for people with high cholesterol levels, to lower their "bad" LDL cholesterol and reduce the risk of a heart attack or stroke.

Although the drug does a great job at lowering the bad cholesterol, there are some major side effects that may make you think twice before taking them. The most common side effect is to your muscles, or more accurately to the mitochondria in the muscles. The pain in the muscles is due to the destruction of mitochondria caused by the statins. Inside the mitochondria, *statins drastically deplete the Coenzyme Q10 levels* as well as produce high levels of super oxide (a free radical) contributing to the destruction of mitochondria that leads to skeletal muscle side effects such as muscle weakness, cramping, soreness and fatigue. In rare cases statins in combination with other drugs can cause rapid muscle breakdown that can lead to liver damage, kidney failure and even death (rhabdomyolysis).

Lack of Coenzyme Q10 completely sabotages the electron transport chain cascade function causing free radical damage and finally destroying the ability to make ATP.

No energy means no muscle function.

Multiple Sclerosis or something else?

I had not seen 45-year-old Donny for at least 2 years. When he showed up at my office on crutches, I thought for a second that he had had some form of surgery or been in a car accident. But when he described his history of why he was using crutches, I had a deep sick feeling that this was not some ordinary muscle weakness that would pass with some rest and good food.

Donny was extremely upset when he told me his story that started around 4 weeks prior to his visit. He had been doing really well since I helped him with his food allergies and digestive problems. He was living a healthful life;

he was sleeping well and was very active physically, doing yoga twice a week and going on 15-mile bike rides three times a week. He was very happy with his career and enjoying his family and intimate life.

Nothing was stressing him out except for the fact that almost overnight, both of his legs stopped working. He had difficulty getting up out of a chair without pain and stiffness. At first, he thought it may have been due to his long bike ride, so he laid off exercising for a few days. But his symptoms kept on getting worse. Within a couple of weeks, he was walking with a weird shuffle because he had trouble lifting his legs off the ground. Running across the street was impossible and he had no strength in his legs to climb up a flight of stairs. Within 4 weeks of the first sign of both of his thighs and lower leg muscles aching, he was unable to drive anymore and felt like a paraplegic.

His primary internist sent him to a neurologist, a specialist who said he needed to get more diagnostic tests to rule out two diseases: MS (multiple sclerosis) and ALS (Amyotrophic Lateral Sclerosis or Lou Gehrig's disease). They sent him home with a set of crutches so he can at least get around his house. Both conditions had very poor prognosis, but ALS scared him immensely because he knew it was basically a "death sentence."

While he was waiting for the MRI/CT scan results, he decided to come in to my office to find out what may be wrong with him.

His fear was as palpable as he said to me, "Dr. Bennett, my legs are dead. My legs are dead! I can't move them anymore. Can you help me?"

I have worked with many neurological conditions including MS and ALS, so I wasn't afraid to dive right in. Since I hadn't seen him for a while, I took a detailed history of the last couple of years and gave a thorough physical exam. I started my routine muscle testing evaluation and immediately I felt that something was just not adding up. In my experience, after taking care of many neurological patients, I found that majority of my MS and ALS patients have had indications of infections, especially viral infections. But Donny was completely clean. He didn't have any other symptoms leading to hidden infections or leading to either one of the neurological conditions.

But his legs were not functioning, he couldn't walk. Could it be heavy metal poisoning? Does he have some genetic disease that just wanted to come out now?

So I started digging deeper being the medical detective that I am. I asked Donny if he had any blood tests from the past 2 years and he was able to pull them right up onto his iPhone. I love modern technology!

I compared two of his routine blood tests ordered by his internist and noted that, over a year ago, he had a high total cholesterol level around 239 mg/dL and the bad LDL cholesterol was definitely in the high range of 169 mg/dL (you want the LDL to be at least lower than 130, but <100 is optimal).

But that was not what really caught my eye. It was the most recent blood test results he had a couple of months before he came into my office that day. It revealed that his total cholesterol and LDL were much lower, 175 and 87 respectively which was in the normal healthy range.

Heavy Metals, Chemicals, Etc. Found in Water

Water is crucial in the metabolism and breakdown of fats in our body as well as shuttling away toxins to be excreted out of the body. With pure unadulterated water, our cellular processes are optimized and this naturally reduces oxidative stress. To reduce the amount of toxic chemicals, heavy metals and other particulates in your tap water, I highly recommend you invest in a carbon-filtered, reverse osmosis water purifier for your home tap. A good reverse-osmosis unit will eliminate chlorine and fluoride, and get out the trace amounts of heavy metals and toxic elements like lead (which is harming the people of Flint, Michigan), arsenic and uranium that are allowed in most municipal tap systems in minute amounts. I don't believe we should be drinking any levels of toxic metals or nuclear waste material. Pure water is best!

Water purification system will leave your water pure and clean and better tasting, and it will supercharge your mitochondrial health. An under-the-counter purifier is a great option for homeowners.

For renters, you can often find a reverse osmosis water purifier that you install directly onto your tap, which will sit on your counter. While a small amount of these chemicals might not seem that harmful, having them build up in your system over time can really do quite a number to your body's ability to produce energy. It's spending way too much time and valuable nutrients trying to cope with and eliminate these metals and chemicals instead of spending energy for the fun stuff you do in life!

If you are interested in finding more information on how to clean up your water, check out Chapter 2 of my book *The 7-Day Allergy Makeover*.

BPA and Plastics

BPA, or Bisphenol A, is a chemical that many of us are only now becoming aware of, and its risks are more evident each day. BPA is found in many plastics and teeth sealants and was even used in baby bottles in the United States until recently. BPA is outlawed in the European Union and Canada. Above all, products with BPA shouldn't be heated (since it can leach into the food you eat or the heated container) and any BPA products with cracks should be immediately thrown away. BPA acts like a hormone and is especially harmful in childhood development. But that doesn't mean it doesn't also do damage to our mighty mitos!

As Chris Meletis, ND writes in his article, "Mitochondria Resuscitation: The Key to Healing Every Disease":

> "In one animal study published in January 2013, exposure to BPA, a toxin commonly found in some plastics, food storage containers, liners of metal cans and

> cash register receipts, causes mitochondrial defects in beta cells [found in the pancreas that excrete insulin]. These defects included depletion of ATP, loss of mitochondrial mass and membrane potential, and alterations in expression of genes involved in mitochondrial function and metabolism."

This depletion of ATP that he cites [7] means less energy for you. The "alterations in expression of genes" means what it sounds like: BPA causes mutations to the DNA of mitochondria, hurting their ability to function. If you're going to use plastics, like plastic bottles of water, it's especially important to make sure they are BPA-free. However, I recommend avoiding plastics altogether for optimum mitochondrial health, not to mention the impact you'll be having on planet earth!

CHAPTER 7

Other Contributors to Mitochondrial Dysfunction

FREE RADICALS FROM FOOD AND AIR aren't the only contributors to poor mitochondrial health. There are several other ways that we can damage our mitochondria, sometimes, even when we think we're doing something that's good for us! These include:

- Over-exercising
- Sedentary lifestyles
- Overeating
- Microbial Infections
- Low oxygenation and poor breathing
- Emotional and psychological stress
- UV Exposure

Over-Exercising

Later in this book, I'll explain exactly why exercise, and the right kind of exercise is so vital for our mitochondrial health. My background in sports medicine tells me that exercise is absolutely one of the best things we can do for our overall health and mood, not to mention our mighty mitos. Obviously, the more we exercise and build muscle, the more we are building mitochondria-rich cells. And that means more energy production!

Over-exercising, however, can be a detriment to your health and can damage our mitochondria. Recent research by Jeffrey Kreher has focused on what's called Over Training Syndrome or OTS. Normally our bodies can handle the stress that sports or other exercise puts on them. However, when we "overreach," that is exercise too much without adequate rest, we may create a system-wide inflammation that produces "effects on the central nervous system, including depressed mood, central fatigue, and resultant neurohormonal changes."[8] Our performance actually goes down and we can start *losing* muscle instead of gaining it. A 1988 publication by Fridén, Seger, and Ekblom entitled "Sublethal muscle fiber injuries after high-tension anaerobic exercise" reveals that over-exercising causes damage to both cellular *and* mDNA.[9] Even if it doesn't damage the DNA, it can make mitochondria "expand" and so decrease their normal functioning. So yes, even exercise can act to hurt or help our mitochondria.

But there's even more to the story of over-exercising. When you exercise too much, you actually aren't doing aerobic exercise (using oxygen to burn calories) but have started anaerobic exercise. Again, some anaerobic exercise is good for you. The problem is when you do too much. The end result of excessive anaerobic exercise is a buildup of lactic acid, which is an endotoxin (a metabolic waste product). It's that chemical that causes cramps and soreness.

But it also results in increased oxidative stress. Furthermore, when you exercise too long, you deplete your body of the nutrients and antioxidant need to feed the mitochondria, you dehydrate your cells (we need water to produce energy!), and ultimately can't remove your body's other waste products quickly enough to prevent oxidative stress.

You're using all your energy, valuable nutrients and enzymes to move faster and train harder instead of performing vital bodily functions that you need daily for ultimate wellness. Overtraining is a mitochondrial killer.

Sedentary Lifestyles

A sedentary lifestyle on the other hand, or what could otherwise be described as the couch potato syndrome, obviously contributes to our poor mitochondrial health. Why? Because not using or moving the body will slow down our metabolism (the rate at which our mitochondria can process food and oxygen) and make our muscles degenerate (muscles are the storehouses of mitochondria). Slower metabolism means a slower rate of clearing out the free radicals that contribute to oxidative stress and mitochondrial DNA mutations. See if you can increase your chances to move every day, whether it's using a stand up desk, walking a few extra blocks after lunch, dancing at home while cooking, gardening or hiking that trail after work. For anyone medically able, a simple walk is a wonderful way to get your metabolism up and your mitochondria reviving.

Overeating

Overeating often goes with sedentary lifestyles. Let's face it, when we feel sedentary, it's often easy to reach for a jolt of energy (usually something sweet) because we aren't getting natural energy from exercise and a diet filled with natural, whole foods. Remember, all foods produce free radicals in some measure through their natural breakdown into energy. That's why it's important not to overeat, since we are creating a higher content of free radicals in the body simply by ingesting more food. Poor food choices that cause excess free radicals are fried foods (which are high in hydrogenated oils), processed meat, alcohol (which *isn't* a food, despite what some of your friends might tell you!), and anything with preservatives (most processed foods).

Another good reason not to overeat is due to the fact that our digestive organs do not have the ability to excrete endless amounts of digestive juices and enzymes to break the over abundant amounts of carbohydrates, fats and proteins taken in at one meal. Not only does this lead to indigestion and reflux, but undigested food travels further down into the small intestines and lower bowels where bacteria can feed on the wasted meal causing gas, bloating, inflammation, pain, cramping, diarrhea and/or constipation. Millions of people walk around with these irritable bowel symptoms that lead to a medical condition called small intestine bacterial overgrowth (SIBO).

SIBO is detrimental to mitochondrial health not only due

to the bacterial overgrowth and inflammation but also from the overproduction of methane and hydrogen sulfide gas by the bacterial fermentation! Gas is what causes the bloating, pain and even mood disorders such as anxiety and irritability.

There is also an interesting correlation between excess bacterial fermentation toxins (D-lactic acid, methane and hydrogen sulfide gas) and disorders such as fibromyalgia, chronic fatigue syndrome and vitamin B12 depletion (all relating to mitochondrial health). The more hydrogen sulfide gas you have, the more pain and Vitamin B12 depletion you have. This was a clinical pearl described by Dr. Alex Vasquez at the 2013 International Conference on Human Nutrition and Functional Medicine. I too have also suspected these correlations after helping countless IBS/SIBO patients and believe that there are definitely clinical findings showing the links between bacterial fermentation, chronic disease and mitochondrial dysfunction.

In Step Three of optimizing mitochondrial health, I will go into what foods to eat and what foods to avoid so you can minimize inflammation, bacterial overgrowth and the dreaded flatulence!

Microbial Infections

Microbial infections, such as bacterial, viral or fungal, trigger mitochondrial dysfunction and are difficult to control once you've been afflicted.[10] There are some obligate intracellular organisms such as the Chlamydophila Pneumonia (previously known as Chlamydia Pneumonia) that infect inside the host cells to leach ATP because the bacteria is unable to make its own energy and will destroy the mitochondria.[11] In addition to upper and lower respiratory infections, *C. pneumonia* has been linked to many other chronic illnesses including atherosclerotic cardiovascular disease, chronic obstructive pulmonary disease, multiple sclerosis and Alzheimer's disease.[12]

The first symptomatic complaint of a viral infection, whether it is a cold or flu, is fatigue. People think the virus is causing the weakness, but if you go deeper, this fatigue is caused by the body's inflammatory reaction causing mitochondrial damage; a natural cascade of immune responses to fight against a foreign invader. I treat a great deal of colds and flu illness, but I also find that chronic fatigue syndrome, dementia and fibromyalgia are linked to viruses such as Epstein Barr Virus, Cytomegalovirus, Human Herpes VI virus and more.

Mitochondrial impairment is also the key to fungal infections such as aspergillosis and *candidiasis* associated with chronic sinusitis and gastrointestinal disorders. All infections increase free radical damage to the mitochondrial membrane and affect optimal organelle function.

Low Oxygenation and Breath

Breath is valued by countless wisdom traditions as the source of life and energy. Have you ever noticed that the more stressed you are, the more shallowly you tend to breathe? It's almost impossible to breathe deeply when you're angry, stressed or upset. On the other hand, the calmer and happier you are, the more you tend to take deep, full, and slow breaths. Guess what? Breathing deeply is crucial for proper oxygenation of your cells and mitochondria. Simply put, the more you are able to get oxygen into your lungs, the more it can be utilized to make energy, or ATP. It's not just good for your mood; it's also good for you mitos! Yoga, meditation, and many spiritual practices know this basic fact of life rejuvenates your spirit and gives you more oxygen to work with at the cellular level. We'll take a look at exactly how to improve our breathing hygiene and oxygenation levels in the final chapter of the book.

Emotional and Psychological Stress

Along with breathing properly, stress reduction is essential for mitochondrial health. Fortunately, the two go hand-in-hand. Better breathing equals less stress. At the chemical level, one of the side effects of being stressed, besides not breathing as deeply and therefore not getting enough oxygen, is that you release a number of hormones in the body. The most notorious stress hormone is *cortisol.* Cortisol, especially in higher levels, will actually go to catabolize, or break down your muscles and replace them with fat. That means that higher stress levels, because they produce more cortisol, tend to reduce muscle, which as we know, is where the highest concentration of mitochondria are. Stress is literally a mitochondria killer! Higher stress levels can also interfere with the methylation pathway, which regulates serotonin and dopamine.

But don't be stressed about your stress! In fact, the more we fight our stress, the stronger it becomes. It's a normal part of life and in the next part of the book, we'll look at ways to reduce and move through your stress and troubles.

UV Exposure

Finally, UV rays can do a real number of harm on our mitochondria. As we all know, excessive UV exposure from prolonged time spent in the sun without sunscreen can cause skin cancer over the long run. However, UV rays cause free radical damage and can denature our mitochondria too. Damaged mitochondria means poor levels of energy within the skin cells to clean up cellular mutations. Again, this all leads to skin disease and cancer.

The first step is to avoid the sun between the hours of 11 a.m. and 2 p.m. when the sun is its strongest. Though excessive amounts of UV rays are harmful to your skin, optimal amounts are necessary in the production of vitamin D, an essential nutrient to our health. Usually you only need to be in the sun without sunscreen for 30 minutes a day in order to get the vitamin D you need. If you are in the sun longer than this, apply sunscreen with an active ingredient of zinc oxide. Another option is titanium dioxide, but note that some people are sensitive to it. Choose one that has an SPF between 21 and 50 along with other natural ingredients. Typical sunscreens with other active chemical ingredients could cause free radical damage and may even be carcinogenic.

Avoiding all of these toxins, sources of free radicals, and harmful lifestyle choices will go a long way toward empowering your mitochondria and getting your health back on track.

CHAPTER 8

Step Two: Purging of Damaged Mitochondria

OVER THE COURSE OF OUR lives, we will all produce damaged and harmful mitochondria. No matter what we do, we can never avoid all UV rays, smog, chemicals in our water and other harmful substances. Even when we eat, we will always produce free radicals as a natural result of metabolism and the Krebs cycle. But there is a great way for purging; getting rid of damaged, harmful mitochondria (*mitophagy*). It's called **exercise**.

Now if you're worried about 2-hour long gym trips and bench pressing three hundred pounds, don't worry. That's not at all what I'm talking about. In fact, as you just learned, overtraining can be just as detrimental to our mitochondrial health as sitting on a couch zoning out. In order to get the most out of our exercise, we should exercise in a way that is best adapted for our mighty mitos. And when we do that, we are accomplishing two things: increasing oxygenation and sending them the message to "cull," or eliminate, harmful and damaged parts of the mitochondria. When that happens, they can undergo what's called "fission" and combine with other healthy mitochondria. The end result is less damaged mitochondria and more and stronger healthy ones. Not only that, but by building new muscles, you increase your "mitochondria to mass" ratio. More mitochondria are ready to go to work for your body.

So what kind of exercise plan works best to purge or cull our damaged mitochondria?

It's what I call *Burst to Boost™ training* (for more information, check the Resources section in the back of the book) or what others refer to as *interval training*. Bursting exercises will boost your mighty mitos! This is a moderate form of exercise that involves alternating periods of medium to high-intensity training with low intensity movements. Best of all, workouts only need to last about 20 minutes. That's right! We only need about 20 minutes, just two to three times a week, for optimal mitochondrial health.

It should be said to check with a medical doctor before undergoing an exercise program, so the following are only general guidelines.

How Does the Burst To Boost™ Program Work?

If you are currently able to walk, instead of taking a 20-minute walk, see if you can walk for 2 to 3 minutes and then add 1 minute of jogging at a level that you would feel is "intense." If 2 to 3 minutes of boost and 1 minute of high-intensity burst feels too difficult, no matter what your level seem to be, you can lengthen your lower-intensity "boosts" and then just do 1 minute of higher intensity "bursting."

The key is to really listen to your body and your fitness level.

Just start by alternating lower level and higher level intensity activities. No matter how long of a "rest" period you need in between, you are exercising in a way that helps build muscle and improve mitochondrial health. There are no wrong ways to do this unless you are overexerting yourself and pushing your limits or not listening to your body. You could even walk for 1 minute and then rest for 3 minutes if that is the conditioning level you are at. That's not a problem. That's *Burst to Boost*™ training for *your* body.

If you can jog comfortably, instead of doing your normal jog at a steady rate for 1 hour, the idea would be to jog at your normal rate for two to 3 minutes and then sprint at perhaps 70 to 80 percent of your maximum rate for 1 minute. After that, you can return to the normal rate of jogging, giving your body a short "rest." After each 2-to-3-minute period of rest, you might increase your "burst" period's intensity, so that you sprint at 85 percent of your maximum rate during the next minute. Don't forget to rest and, if you need more time, no worries. Keep increasing the speed of the burst period, especially as your body over the coming weeks and months adapts to new levels of fitness. You eventually want to get to a 1:1 ratio of burst and rest. One minute sprinting and 1 minute of walking or light jogging.

My personal favorite *Burst to Boost*™ exercise is where I run on a treadmill at the steepest incline for 1 minute as fast as I can, walk for 1 minute, and then immediately go into speed jump roping. Please note, my exercise technique is for someone who is highly conditioned or a seasoned athlete. If this type of training is new to you, or you're recovering from a surgery, had an athletic injury or have a chronic illness such as chronic fatigue syndrome, fibromyalgia, or any other inflammatory disorders, you need to start very slow, with very light exercise patterns. One of my patients who has a muscular dystrophy disorder can stay on his elliptical exercise machine for only 5 minutes before he tires out. If that is where you are at, listen to your body and don't push it. It may take months for you to get to a good level of training; be patient. Don't forget to talk to your primary health care provider and get his or her approval first.

Of course, jogging isn't the only activity you can do. I've found that stationary bikes are often the type of exercise that my patients like best. It's easy to cycle casually for a bit and then go all-out for a bit. That "rest" period feels so good afterward. There are lots of forms of exercise that are naturally suited to this type of *Burst to Boost*™ training, including CrossFit, climbing stairs, and certain gym classes. However, for best results, try to work toward timing each burst period at equal, alternating intervals and don't do the exercise more than 20–30 minutes.

On off days, you are welcome to do exercises that involve stretching and light movements such as Qigong, tai chi, Pilates and restorative yoga. These light movement techniques help move your lymph and blood flow, improve your flexibility and tone, and boost your energy and life force (qi or chi). I have been practicing Qigong for a long time and I train my body's energy level with a special form called Animal Qigong—mainly the crane, deer, turtle and tiger exercises (for more information, check the Resources section in the back of the book.)

Why Does Burst To Boost™ Work?

Now that you know what *kind* of exercise is best for your mighty mitos, you might want to know exactly why it works at the mitochondrial level. Put simply, exercise helps with mitochondrial **fission**, in which a mitochondrion will break off the bad parts of the organelle so that it can fuse with another mitochondrion, known as mitochondrial **fusion**. By doing *Burst to Boost*™ exercise you are literally destroying the bad parts of your mitochondria and allowing healthy parts to come together.

Next, because you are moving between intense states and "resting" states, considered at about a 50 percent drop in intensity, you increase the amount of oxygen that can get delivered to the muscles. Doing too intense of an exercise for too long results in *anaerobic* exercise, which only builds lactic acid in the muscles and doesn't help increase oxygen flow. Also, performing at a slower but steady rate like long walks or a hike doesn't help increase this oxygen delivery as quickly for such a short period of time.

Better yet, with the interval training your body adapts and quickly improves your fitness level so that you have less chance of damaging yourself and your mitochondria, compared to a six-mile jog or continuous, super-intense workouts.

By using *Burst to Boost*™ exercises you also enhance the ability of the muscles to use fat as an energy source through increased oxygen utilization. I don't mean burning fat during the exercise per se, but actually after the workout. In fact, you can burn fat up to 48 hours after, even if you are not working out. This very important post-burst training phase is called the "after-burn." Your body keeps up its ability to use oxygen to burn fat. This means that mitochondria are supercharged after exercising, since with more oxygen they can use fat as an energy source. Not to mention that your overall metabolism increases, which makes the production of ATP and other chemicals processes, including the elimination of toxins and those harmful free radicals, occur more quickly and efficiently.

That's why I recommend you train only two to three times of interval training a week; put it simply, you must give the body the rest it needs to purge the damaged mitochondria, strengthen your muscles to upregulate oxygen utilization and increase your fat metabolism.

Examples of Burst to Boost Exercises (1 minute on at 70–85 percent of maximum physical exertion/1 minute off for a total of 10 repetitions):

- Running
- Jump-roping
- Climbing stairs
- Elliptical, StairMaster or VersaClimber machine
- Nordic walking
- Biking (Road, mountain or stationary)
- Dancing
- Calisthenics (sit ups, pull ups, jumping jacks as fast as you can for 1 minute, then 1 minute off, at rest)
- Swimming laps
- Kick boxing
- Wrestling

What a win-win situation--who doesn't want to have an exercise plan that is just 1 hour a week (20 minutes a day, three times a week) that burns fat when you're not even exercising, including during your sleep? We need to ingest healthy fats for energy but over abundance of stored fat is considered an inflammatory organ; it can be detrimental to your heart, brain and the mitochondria.

Please note, I want to be clear about something here: I am not saying that you will become this amazing healthy slim person if you only move 1 hour a week. What I am saying is that to improve mitochondrial health and improve your ability to burn fat, you only need three sessions of high intensity exercise per week. I recommend that you include whole body exercises to prevent bone and muscle loss, for your flexibility and structural health. Movement is key to life!

Examples of Whole Body Exercises:

- Tai chi
- Nordic walking
- Egoscue exercises
- Active and passive stretching
- Qigong
- Snorkeling
- Stand up paddling
- Skateboarding
- Beach walking
- Biking with family

- Tennis, racquetball or paddle tennis
- Ping pong or badminton
- Playing softball with your kids

Let's not forget that exercise also increases blood flow to the organs so that antioxidants and vital nutrients get to them more efficiently and quickly. This is where nutrigenomics is crucial. The quicker we can deliver vital nutrients to our cells, including our DNA, the more we can control our genes. And that means turning on and off the genes that control diseases from Alzheimer's to depression to cancer.

And if you needed one last reason to start interval training today (besides the fact that's it quick, easy, and free), here it is: it will leave you in a good mood. Almost nothing is better for mood than exercise. Exercise lowers cortisol levels, and as said earlier, this stress hormone destroys muscles. So let's get moving every day, no matter how much.

Other than doing Burst to Boost™ exercises, I highly recommend walking for 15 minutes after every meal, especially for people on the diabetes spectrum or with blood sugar issues. Walking will help bring down your blood glucose optimally regulating your energy levels.

But no matter what, as said earlier, incorporate some physical activity each day. It doesn't need to be strenuous either, just start moving! Here are some more ideas:

- Start gardening
- Join your kids in their sports or play

- Walk to the store instead of driving
- Explore a new neighborhood on foot
- Hike that trail or go snowshoeing with a friend
- Do a home renovation project involving physical activity
- Park further away from work, so you have a longer walk
- Find a walking buddy or join a walking meet up
- Dance to your favorite song with your pup
- Shoot hoops
- Take the stairs instead of the elevator

Other Ways To Encourage Mitophagy

Along with exercise, there is another way to promote mitophagy and that is through calorie restriction or intermittent fasting. I am not going to go into details about this technique in this book because I believe that it warrants a health practitioner's supervision for optimal results, but let me explain a few points.

Basically scientists have found that calorie restriction and intermittent fasting stimulates mitophagy as well as mitochondrial biogenesis (production) and can delay the aging related degeneration, be cardioprotective and improve a robust lifespan. In another words, calorie restriction and intermittent fasting trigger culling or purging of the mitochondrial pool while improving longevity and health of the individual.

In addition to fasting for short bouts to stimulate mitophagy, carbohydrate restriction can be implemented. Calorie restriction and intermittent fasting are not the same as carbohydrate restriction. Calorie restriction and intermittent fasting is when you are eating 30 percent or lower amounts of calories per day or fast for 12 to 15 hours at a time, respectively. Carbohydrate restriction is when you remove most of the simple and complex carbohydrate foods out of the diet and counting calories is not as important; in fact, you may actually find that the diet can be quite high in calories, mostly from fats and proteins.

Carbohydrate restriction can be considered a ketogenic diet, where one eats moderate to high levels of fats, moderate levels of protein and very low calories of carbohydrates. One of the benefits of a ketogenic diet or low carbohydrate diet is that it causes lipolysis (fat breakdown) that has a positive effect in the production of ketone bodies such as beta hydroxybutyrate, which stimulate the electron transport chain and mitochondrial function in addition to purging of damaged mitochondria.

This type of diet has become very popular for rapid weight and fat loss. And again, I can't reiterate enough that medical supervision is highly recommended, here is a case study explaining my concerns.

Danger Zone

A female patient came into my office feeling gravely ill after being on a high fat diet for 4 days. She said she used the exact steps recommended in a very popular ketogenic dieting book. She felt sick to her stomach, been having six or seven bouts of greasy diarrhea for days, and was very dizzy and lightheaded.

She was excited that she was losing weight but she knew deep down something was seriously wrong with her.

I observed that she was breathing more rapidly than normal. After checking her pulse and blood pressure, I told her to stop the diet right away. I told her that the high fats she was consuming was the cause of the diarrhea, that I suspected that the ketone levels were stressing out her kidneys and that my biggest concern was that she may be in a high ketoacidosis state where her body was having trouble getting rid of the excess ketones. This was causing a metabolic condition that is considered a medical emergency! I immediately stopped my exam and sent her directly to the emergency room.

Sure enough, the emergency room doctors told her the exact same information as I did, gave her an IV drip to restore her electrolyte balance and kept her in the emergency room until her blood markers were back to normal before she was released.

Not everyone can use this method for purging damaged mitochondria, weight control and other health benefits. This is why I recommend the Burst to Boost™ exercise regimen to purge damaged mitochondria rather than calorie restriction, intermittent fasting, or the ketogenic diet.

CHAPTER 9

Step Three: Protect and Produce Healthy Mighty Mitos

NOW THAT YOU'VE LEARNED HOW to prevent mitochondria impairment by reducing free radical damage and how to use interval training as a method to purge bad mitochondria and improve the process of mitochondrial fusion, we're going to look at how to produce new mighty mitos (*mitochondrial biogenesis*) and protect them.

Nutrition is where we begin and perhaps the most important thing you'll learn in this book. The fastest way to improve your energy level and mitochondria's ability to make more ATP is through the Mighty Mito Nutrition Plan.

This includes five steps:

1. Eliminate the obvious
2. Remove seven types of foods that trigger inflammation and mitochondrial stress
3. Eat natural, colorful, organic foods to keep mitos happy

4. Supplement with mitochondrial enhancing nutrients for the extra boost
5. Eat Oxygenating Pudding to spark up your life

CHAPTER 10

Mighty Mito Nutrition Plan Step1

Eliminate the Obvious

WHAT COULD BE MORE IMPORTANT than the nutrients you put in your body for helping your mitochondria produce energy, helping your body fight off free radicals, and contributing to the muscle building and a healthy spectrum of weight for you to operate at? There are certain foods, liquids and chemicals in food that trigger high levels of free radicals and cause mitochondrial dysfunction and, for this reason, you should avoid them altogether. These are

- Alcohol including wine and beer
- GMO foods (glyphosate-drenched soy, corn, canola and wheat and now we have GMO salmon!)
- Pesticide, herbicide or fungicide contaminated fruits, veggies, grains, nuts/seeds, etc.
- Rancid or denatured oils (cooked at too high of a temperature)

- Fried foods, especially those made with denatured oils
- Trans fats or partially hydrogenated fats (found in margarine and in many processed junk food)
- Processed meats (sausage, bacon, salami high in nitrates)
- Sulfite drenched dried fruits (apricots, raisins, mangos, etc.)
- Fungal contaminated foods (cheese, dried fruits, peanuts, cashews and pistachios, packaged snacks, bruised and overripe fruits, old veggies)
- Food coloring and food additives
- Artificial sweeteners (acesulfame, aspartame, saccharin, sucralose)

CHAPTER 11

Mighty Mito Nutrition Plan Step 2

Remove 7 Types of Foods that Trigger Inflammation and Harm Your Mitochondria

THERE ARE MANY FOODS THAT we eat which are considered healthy and full of nutrients, but we have no idea that there are substances in these foods that may be adding to your fatigue and overall inflammation in the body because it's hidden deep inside your gut.

For this reason, I'm going to ask you to follow my Ultimate Wellness Food Checklist (UWFC) for the next 6 weeks (the complete list is in the Appendices and Resources section.) See how your body responds. I will be surprised if you don't start to feel better with more energy, better sleep, and clearer brain. I've counseled hundreds of patients who didn't even realize that their bloating and gas were due to a dairy or gluten intolerance that also made them feel tired, cranky and inflamed!

Here are the benefits to following the Ultimate Wellness Food Checklist:

1. Protect, build and support your mitochondria
2. Boost your energy levels
3. Sharpen your mental focus and clarity
4. Reduce free radical damage, inflammation and pain
5. Improve your gut health and reduce your IBS/SIBO symptoms, especially gas and bloating
6. Help build a strong and powerful vibrant body

From my clinical experience after seeing countless patients with chronic diseases as well as from my own experience after a TBI, head and neck injury, the seven food types below contribute to intestinal bacterial overgrowth, trigger local and systemic inflammation, cause excess free radical damage, and ultimately impair mitochondrial function.

The seven food types to be avoided are

- **Dairy** (in cheese, milk, yogurt, ice cream; Note, eggs are not considered a dairy product, they come from a chicken. So eat unless you are allergic to them)
- **Gluten** (in wheat, barley, rye, kamut (Khorasan wheat or Pharaoh grain), spelt and even soy sauce!)
- **Sugar** and processed sweeteners (white sugar, brown sugar, honey, agave, sucrose)
- **Alcohol** including polyols (sorbitol, xylitol, mannitol)
- **Beans** and legumes (including soy, peanuts and cashews)
- **Vegetables** high in fermentable carbohydrates
- **Fruits** high in fermentable carbohydrates

Some of these foods such as dairy and gluten grains can be considered as food sensitivities and allergies that I've detailed in my book *The 7-Day Allergy Makeover*. But here, I am addressing other valid reasons why it's necessary to remove the dairy and gluten grains for optimal health.

In short, these seven specific foods types can trigger inflammation in the gut and as well as inflammation systemically, and slow down your metabolism and levels of ATP production. In addition, there are fruits and vegetable foods I will be asking you to eliminate that may cause some emotional distress because they are considered rich in vitamins and antioxidants, such as avocados, apples, broccoli, asparagus, bell peppers and mushrooms.

In fact, I will ask you to remove thirty-six fruits and vegetables for the first 6 weeks to improve your energy, brain and physical power. But don't worry, I give you a much longer list of foods to eat and enjoy. Most of my patients say that after a couple of weeks of following the UWFC, they don't miss the sugar, fruits and veggies on the "do not eat list" anymore because they found the foods that were allowed were full of taste and substance.

Now I have created this comprehensive list after many years of research and clinical experience, and it might seem like a lot to eliminate at first. I understand how hard it can be if you've been accustomed to eating pizza, pasta, donuts and other foods that contain one or even multiple of these ingredients. But you'll soon see there are still plenty of amazing- tasting, beautiful foods left for you to explore. There is a Mighty Mito Recipes bonus section at the end of this book, including many mouth-watering meals I make regularly. And, best of all, with this new Mighty Mito nutrition program, you don't have to count calories or limit portions, even if weight loss is one of your goals. You simply focus on eating more natural, colorful, organic foods.

Why Can't I Eat These Foods?

Before I go into each of the seven food types, let me first explain why these foods are so important to avoid. All seven types of foods on my "Don't Eat list" have **natural sugars (fermentable carbohydrates)** in them that are difficult to digest and absorb; our digestive organs do not make the proper enzymes to break them down completely. These undigested or malabsorbed sugars travels deeper into the intestinal tract *feeding the bacteria* which leads to *fermentation.*

By using fermentation, you are turning your body into a "brew house" temporarily, relying on the trillions of bacteria in your gut to do the work of extracting nutrients and energy from food. Now, that's not bad in itself. Those gut bacteria are there for a reason and they help us get the most out of a variety of foods every day. But the problem I find is that most of us walk around with an imbalance within our microbiome (gut bioflora), partly due to eating the wrong foods.

Eating excessive amounts of fermentable carbohydrates combined with excess bacteria in the gut produces bloating, flatulence, abdominal pain, diarrhea, constipation, and a host of other symptoms leading to inflammation in the body, free radicals damage to major organs and mitochondrial dysfunction.

Some of these fermentable carbohydrates (fructans, galactans, lactose, and polyols) can also cause an increase in the osmotic gradient where water rapidly is drawn into the intestines causing explosive diarrhea. As mentioned earlier in chapter 6, when these bad bacteria get out of control, they can produce an excess of chemicals that are actually extremely harmful to our mitochondrial health as well, including d-lactic acid and hydrogen sulfide gas in particular.

All seven types of foods I ask you to remove have fermentable carbohydrates that promote bacterial fermentation, which leads to SIBO (small intestinal bacterial overgrowth), inflammation, irritable bowel symptoms and mitochondrial impairment.

While some people are less sensitive to fermentation—especially if they have built stronger microbiome (gut bacteria)—try eliminating some fermentable carbohydrates from your diet and then slowly reintroduce them to see how you feel. You might find, for instance, that you do fine eating broccoli and Brussels sprouts (which are fermentable carbs), but that it's harder to eat black beans. Listen to your body. These foods are generally very good for you and *should* be eaten, if you don't have a sensitivity to them. After eliminating a great majority of fermentable carbs, you can then come back to the ones you enjoy and that your body enjoys!

By temporarily cutting down on fermentable carbohydrates you are supporting your mitochondria and will feel healthier, lighter and more energetic as your gut returns to its normal, optimal state.

One of my favorite quotes is by Hippocrates, the Father of Medicine, "All disease begins in the gut."

To improve the health of your mitochondria and prevent disease, it's imperative to reduce bowel inflammation and toxicity, and you can achieve it by removing the following seven types of food.

Dairy

Dairy (animal milk products from cow, sheep, goat, buffalo, camel and yak) naturally has a fermentable carbohydrate called lactose. Ingesting lactose can produce symptoms like bloating, gas, and diarrhea and can cause inflammatory reactions that can contribute to headaches, joint pain, acne and much more. After the age of 5 years old, our digestive system greatly reduces the amount of lactase, the enzyme needed to break down lactose. Lactose-free milk is not recommended either; there are other ingredients such as casein and whey that humans have trouble digesting and can also contribute to inflammation in the gut. Alternative milks and cheese made from almond, coconut, hemp and rice are great replacements.

Gluten

Gluten grains (wheat, rye, barley, kamut, spelt, triticale) are very taxing for most people to process. Even if you don't feel the brain fog, food coma, or tiredness that many people do, it can cause low level inflammation so that your body will be spending its extra energy (ATP) fighting that instead of working toward optimal health. The *fermentable carbohydrate* in gluten grains are called fructans, which are strands of fructose joined together, and when ingested, they cause IBS and SIBO symptoms.

Sugar

Sugar is one macronutrient we all need to maintain a healthy blood sugar (glucose) level; it provides the quickest energy source produced through glycolysis and glucose is the main source of substrate that the brain utilizes to make energy. But too much blood sugar (whether from poor diet or even a gluten-free snack) can cause free radical production causing inflammation in your body. High sugar intake also trigger the formation of advanced glycating end products (AGEs) that can cause your blood to become sticky which cause more oxidative stress damaging your blood vessels, heart tissue, and other important organs. Remember, oxidative stress of any kind will cause damage to your mitochondria. As discussed in the purging of damaged mitochondria, one of strategies to clean up bad mighty mitos is to implement a low carbohydrate diet.

If you have taken a look at my Ultimate Wellness food checklist, you will notice that I recommend healthy carbohydrates for you to enjoy, particularly sweet potatoes, butternut squash, nuts and seeds, and ancient grains such as quinoa and amaranth. Another reason I don't recommend low carbohydrate diet with most of my patients is they all come in with adrenal dysfunction or exhaustion due to the modern stressful life. If you have an underlying adrenal issue and then completely cut out all of the carbohydrates, you will find out pretty quickly that you made a huge mistake; your blood sugar will crash, and you will feel fatigue, lightheaded, dizzy, irritable, and anxious and have insomnia. You will also start craving sugar all day. Who needs that! These are symptoms of low blood sugar called hypoglycemia. This is a medical condition that needs to be addressed by your holistic health care provider.

Natural sugar from real, whole fruit (blueberries, oranges, coconut) is fine. But we want to avoid all sweeteners, natural or artificial, such as white and brown sugar, honey, agave, sucrose, fructose, and xylitol. The exception is stevia, which is made from plant leaves and contains no sugars but still manages to taste sweet as well as erythritol powder which is a low fermentable sugar, made up of a four-carbon alcohol sugar and easy to cook with.

Also, make sure the food or supplements you buy don't have lactose as an added ingredient, which is an animal milk product. You'll be surprised how much lactose is added to our foods, snacks, drinks and even supplements. I had to send back a whole case of supplements because lactose was hidden in as one of the "other ingredients."

Alcohol

Alcohol, which we've covered, should be avoided or at minimum greatly reduced, perhaps one drink a week. If you are going out for a work celebration or to a birthday party, best choice for a festive drink is one that's made from soju (Korean rice vodka), sake, tequila or gluten-free vodka. Of course don't make the drinks with soda pop, margarita mixes or high fructose corn syrup. "Skinny" drinks are best by adding lime juice, bubbly water, a couple of drops of alcohol-free liquid stevia to cut the sour taste and lots of ice to water it down. Stay away from beer and wine since it contains sulfites and mold.

Beans, Legumes and Soy

You should avoid beans, legumes and soy for the first 6 weeks. While they may serve as a protein substitute, especially for vegans and vegetarians, our bodies are not well equipped to deal with large amounts of these foods due to the fact that humans do not produce the appropriate enzymes to break down the tough-to-digest galactans (fermentable carbohydrate) that naturally occur in beans. Eating beans increases inflammation in the gut and more. Everyone has experienced a bad case of gas after a big bowl of chili! Besides being a fermentable carbohydrate, countless people are actually sensitive to soy. When you ingest food that you are sensitive to, it triggers an inflammatory cascade of events and immune factors are released, increasing oxidative stress and free radical damage. People with foods allergies are not energetic people with exuberance and enthusiasm. Fatigue, stomach pain and brain fog are very common symptoms, all very similar to mitochondrial dysfunction.

Instead of beans, try quinoa, amaranth, hemp hearts and chia seeds. These alternatives are very high in protein and are free of fermentable carbohydrates.

Fruits and Vegetables

Finally, fruits and vegetables high in fermentable carbohydrates are the last two to be avoided just for a little while. As discussed earlier, to reduce the inflammation, free radical damage and bacterial overgrowth in the gut, you will need to refrain from eating thirty-six different fruits and vegetables that are considered healthy for the first 6 weeks of your nutrition plan.

Starting the Mighty Mito Nutrition Plan

Eliminating these seven types of foods above can go a long way toward taking the stress off your mitochondria. While the change might seem like a lot at first and you may be asking yourself right now, "What is left for me to eat?" Let me start by saying there's good news! There's a world of meats, vegetables, whole grains, fruits, herbs, nuts and seeds to eat. Don't get rid of your favorite bolognaise spaghetti meal. Replace your pasta with quinoa instead. Free-range chicken and vegetables might seem like a flavorless, boring meal, but you'll be surprised just what tasty gluten- and dairy-free creations are out there online. Plus don't forget the Mighty Mito recipes in the appendices.

My patients realized that by simply getting rid of foods high in fermentable carbohydrates, their symptoms of tiredness, brain fog, blood sugar crashes, muscle and joint pain, even anxiety and depression, lifted within 2 to 4 weeks. Implementing the Mighty Mito nutrition plan is so powerful because it allows your mighty mitos to get turned on to do what they do best: produce energy for vibrant living!

I recommend printing out the Ultimate Wellness Food Checklist or taking a photo of it so it's conveniently available when you go grocery store shopping. Over time, it will become intuitive which foods you can eat during this period and which to avoid. Some examples of foods to avoid that are otherwise very healthful foods include:

- Cabbage
- Beans, soy and lentils

- Onions
- Garlic
- Broccoli
- Brussels sprouts
- Apples and pears
- Peaches, plums and nectarines
- Watermelon

After a few weeks, notice how your body feels without these foods high in fermentable carbohydrates. Are you finding that you're late afternoon "pregnancy" belly is starting to disappear? Your embarrassing gas has gone way down? You wake up with more energy and alertness, as you did when you were in high school? That's the process of giving more power to your mitochondria and its aerobic processes!

Reintroduction of Foods

After 6 weeks (for some people it may be 8 weeks or even 3 months of the UWFC) following step 2 of the Mighty Mito Nutrition Plan, the Ultimate Wellness Food Checklist, you can reintroduce the foods high in fermentable carbohydrates slowly, one at a time, every 4 days and see how you feel after each new food item. A good place to start is to reintroduce the fermentable vegetables on the Do Not Eat List first, then try a fermentable fruit next. If things are going well, how about adding a ¼ cup of lentils in your soup.

Keep a detailed food diary, so you can track your daily food plan, each new food that you reintroduce, and jot down your symptoms. Be mindful of what your body is telling you. Having trouble with your energy in the afternoons again? Are your headaches coming back? Maybe some gas is starting to creep up? How about your aches and pains. Are they coming back, joint by joint? If your symptoms reappear within a couple of days, go back through your food diary and you will be able to identify which specific foods are the triggers.

After treating thousands of patients with chronic fatigue, joint pain, allergies, environmental illnesses, all relating to mitochondrial dysfunction, there is one caveat to the reintroduction phase. I highly recommend you not reintroduce dairy and gluten grains back into your diet, ever. I know this is asking a lot out of you, but from clinical experience, after having logged over 130,000 patient office visits, I found that these two food groups are closely linked to mitochondrial impairment and many of your irritating chronic symptoms and illnesses including chronic fatigue, pain and inflammation, digestive disorders, headaches, fibromyalgia, autoimmune issues, anxiety, insomnia and more!

CHAPTER 12

Mighty Mito Nutrition Plan Step 3

Eat Natural, Colorful, Organic Foods to Keep Your Mitos Happy

SO WHAT DO YOU WANT to eat?

In the coming weeks, you'll be shifting your body to rely more and more on burning fat, protein, and carbohydrates than on burning the simple sugars and fermentable carbs that we normally eat in foods like wheat, dairy, white sugar, beans, and certain fruits and vegetables.

At the most basic level, you'll want to begin eating more and more *real, whole, natural* foods. This means that as much as possible you'll want to eat only the things that nature has provided in their original state: free-range, grass fed/grass-finished meat; wild caught fish; cage-free organic eggs; organic fruits and vegetables; nuts; and seeds. This is how humans have eaten for 2 million years, and it has worked very well for our species! Add in a healthy supply of whole grains, and you have the benefits of the last 10,000 years as well. On my *Ultimate Wellness Food Checklist,* you have a complete list of what foods to eat that are low in fermentable carbohydrates. There are so many more foods you can eat than the ones I ask you to avoid, so enjoy!

Some examples of these foods are

- Free-range beef, pork, chicken, and lamb
- Free-range eggs
- Fish (1 time a week to limit mercury and nuclear waste intake)
- Zucchini
- Arugula
- Amaranth
- Spinach
- Bok choy
- Swiss chard
- Tomatoes
- Blueberries
- Red bell peppers
- Quinoa
- Sweet potatoes
- Almond, coconut, and rice milk
- Walnuts, almonds
- Pumpkin seeds

Again, this is just a *partial* list. You'll see dozens of other great foods on my Ultimate Wellness Food Checklist found in the appendices and resources section.

The Antioxidant Revolution

Now, there's one final thing you'll want to understand in your new eating program as you carry it forward throughout your life. You'll remember that to some degree all food produces free radicals and don't forget the daily barrage of toxins we are exposed to that add to the concentration of free radicals.

As discussed earlier, the damage caused by free radicals is called "oxidative stress."

However, we also mentioned that our body can normally handle a certain amount of free radicals and that's because we have natural defenses against oxidative stress known as endogenous *antioxidants* coursing through our systems, called glutathione, catalase and superoxide dismutase (SOD).

These antioxidants are helpful chemicals that donate their electrons to the free radicals so that these free radicals don't get the chance to attack our cells. That means instead of binding to our mitochondrial DNA for instance, the free radicals will bind to an antioxidant coursing through your system.

Imagine them like magnetic attractors, pulling free radicals away from our cells and into a place where they can be more easily used and disposed of. But when our endogenous antioxidants can't handle the free radicals, especially when there's an overabundance of them, energy production in the mitochondria is compromised, leading us to feel fatiqued, lose focus, and have lower immunity.

Where to get these "magic" antioxidants then?

The truth is they're not a great hidden secret. Vitamins A, C, and E are antioxidants. But there are thousands of antioxidants. In fact, they're in many of the foods you already eat, and some of the tastiest foods on earth, such as salmon, blueberries, and dark chocolate, just to name a few. But the truth is, the standard American diet, filled with things like white bread, processed meats, pizza, spaghetti, soda, donuts, and fast food that are often both nutrient-poor compared to healthier, whole foods and severely lacking in antioxidants. When you eat foods low in antioxidants, you become less capable of processing the free radicals and the production of energy—ATP.

How do you know what foods have lots of antioxidants? Most antioxidant-rich foods follow this simple formula: **they are natural, colorful, and organic.** These nutrient-dense foods are what the Mighty Mito nutrition plan focuses on since they're best for the mitochondria at the nutrigenomic level.

By **natural** we mean foods in the least-processed state possible. Tomatoes you buy in the store or farmer's market. Fresh whole fruit rather than diced fruit cup soaked in sugar. Kale you harvest from your backyard. That doesn't mean you can't eat canned or packaged foods, but you want to make sure that it has *no additives, no preservatives, nothing chemical* about it. So a tomato you buy whole is better than salsa with preservatives and chemicals you can't pronounce. But if your dinner recipe requires organic canned, diced tomatoes, that's okay, as are natural store-bought organic salsas when you don't have time to make it fresh yourself. Just think about eating more farm-to-table style. If you follow that one simple strategy you will already be eating better than 90 percent of the American population!

This brings us to the second point. You want **colorful** foods. Not only will they make your plate and palate happy, but color is often (though not always) a sign of high antioxidant, rich in minerals, and loaded with vitamins. Generally, the deeper the color of the fruit or vegetables, the higher concentration of antioxidants such as carotenoids found in yellow and orange vegetables (squash, bell peppers, sweet potatoes) which are precursors to vitamin A. Remember how your mom or dad told you carrots were good for your eyes? Well, they're also good for your mitochondria!

Green vegetables, especially dark green ones (kale, collard greens, arugula, and various herbs) tend to be high in lutein and zeaxanthin, both antioxidants. Dark green vegetables also create an alkaline (rather than acidic) environment in body, and an alkaline system helps facilitate the electron transport chain, which is one of the steps of the ATP-producing Krebs Cycle. By eating colorful, natural foods, you are delivering, at the nutrigenomic level, the best possible nutrients to your system and providing your body with antioxidants to fight off free radicals and to turn off bad genes!

But don't be fooled by all those colors. Unless it's **organic**, the companies may have used artificial dyes in order to color their fruits, vegetables, and meats. That brings us to our final point. We already learned that pesticides and herbicides can contribute to our oxidative stress levels. But they can also lead to low level inflammation, slowing our metabolism and weakening our system. Organic is *not* a fad. Our species has been doing it for 2 million years. It's only in the last century or so that we have strayed from this healthier and more mitochondria-empowering way of eating.

Natural, colorful, organic. You can even remember the formula this way: *N.C.O. for the Mighty Mito.*

To give you just a partial list of some of the most antioxidant-rich foods that are low in or free of fermentable carbs:

- Brazil nuts
- Walnuts
- Kale
- Wild Salmon (astaxanthin)
- Pecans
- Blueberries
- Cranberries
- Spinach
- Eggplant
- Hazelnuts
- Ginger
- Cloves
- Pomegranates
- Collard greens
- Rainbow chard
- Sweet potatoes
- Carrots
- Fresh herbs (thyme, rosemary, sage, cilantro)

By eating these foods, you'll be able to easily fight free radical damage through the all-powerful antioxidants they contain.

Where to Start?

There's also a lot we can do in the *way* we eat in order to improve our mitochondrial health through nutrient absorption and eating the right kinds of foods.

First, unless you are using the intermittent fasting technique for a short period to purge your damaged mitochondria (mitophagy) then I recommend you eat a small amount of protein at every meal especially during breakfast. A couple of eggs is fine or some organic chicken sausages with the rest of your breakfast. What this does is help stimulate muscle protein synthesis, as well as continue to strengthen and repair muscle, which as we know is where our mitochondria are most highly concentrated.

Protein, as we are beginning to fully appreciate, helps not just with muscle but with a range of factors in your health, including neurotransmitter ("feel good" brain chemicals) production, immune boosters, enzyme production, detoxification, and the methylation pathway. Eating more protein often leaves you more full and less in need of carbohydrates. By eating fewer carbs, our bodies start to burn fat, instead of sugar for energy, which is also much easier on the mitochondria. While you don't need to eliminate carbohydrates, shifting toward healthy ones in vegetables and whole grains while boosting your protein intake will go a long way toward helping your mitochondria thrive.

When you do eat protein, just know that not all proteins are created equal. What is the best protein, you might ask? The winner is eggs, with 48 percent of the protein in eggs being utilized by the body. If you are sensitive to eggs, you have to find other options. Second, foods like meat, fish and poultry come in at about 32 percent absorption. Soy proteins are only 18 percent utilized by the body, which is yet another reason to avoid it.

Second, if you find yourself getting gassy and bloated after eating animal protein, it may be from not excreting enough digestive juices to break down the protein. If so, I suggest taking a digestive enzyme that includes betaine HCL and ox bile with the enzymes. These supplements will optimize digestion and help balance the gut bacteria and microbiome.

Finally, be sure to eat slowly and chew your food at least twenty times with each bite. You will extract more valuable nutrients from you food, the more mindfully and carefully you chew. Don't rush. Put your fork down between bites, even if the food is so tasty you're tempted to shovel it into your mouth. We've all been there!

I don't remember who taught me this quote many years ago, but it stuck with me: "Drink your food and chew your liquids." Basically it's saying to chew your food until it turns into liquid, so you can drink it and chew your liquids to activate your intestinal juices.

And don't forget to savor and enjoy your food fully; you're powering up your mighty mitochondria for vibrant health!

CHAPTER 13

Mighty Mito Nutrition Plan Step 4

Supplement with Mitochondrial-Enhancing Nutrients for the Extra Boost

SIMPLY IMPROVING YOUR DIET CAN go a long way toward healing your body. However, you can get a wide variety of antioxidants from supplements that you might not be getting in your normal diet. In this section, I'm going to share with you twenty nutrients vital for optimal mitochondrial health. In fact, these are the exact same nutrients that helped restore my brain and body after my terrible injury. In addition, I believe they have helped slows down my aging process!

You will be able to find many of the following nutrients in a combination mitochondrial formula, but some you may need to take individually. In choosing what to take, I recommend going to a natural pharmacy, get them from your health care practitioner, or find companies that don't source their ingredients from China or India. Both countries have very polluted waters and may affect the quality of the raw ingredients.

Discuss all health implications with your doctor before beginning any new supplement regimen and find out which nutrients fit your specific health needs.

Quality Multi-Vitamin and Mineral Formula

Along with the specific antioxidants and mitochondria-enhancing supplements listed below, it's always a good idea to take a multi-vitamin/multi-mineral formula, full of essential or *orthomolecular nutrients* including: vitamins A, B complex, C, D, E and K, and essential minerals such as magnesium, calcium, zinc, selenium, chromium, potassium and more.

Orthomolecular is such a big scientific word but basically, it's referring to nutrients that are truly essential for the human body. Orthomolecular nutrients support the body to thrive optimally, prevent cellular and mitochondrial damage, and prevent early aging and disease.

Our body cannot produce any of these nutrients on its own. These are nutrients that we have to have daily; otherwise, our cells start to deteriorate causing mitochondrial stress and early aging. Therefore, a multi-vitamin/mineral is a good start for mitochondrial and overall health.

On another note, before purchasing your vitamin/mineral formula, find one that has natural folates instead of "folic acid" which is synthetic. Synthetic "folic acid" is not the same as "folates." Studies are now showing that folic acid that's been added for good reason (to processed foods such as bread and pasta to fortify the B vitamins stripped by the manufacturing process), can be harmful to your body, mitochondria and methylation pathways. It's particularly important for pregnant mothers, so make sure your prenatal vitamins have natural folates, rather than folic acid.

In my practice, I order a blood or saliva test that looks for methyl genetic defects including the MTHFR (methylenetetrahydrofolate reductase), a genetic mutation that affects optimal folate and methylation metabolism. Many people have this genetic defect (35–75 percent of the US population), and if so, a special type of folate called L-5-MTHF (5-methyltetrahydofolate) is recommended. Ask your doctor to test you for this genetic mutation and, depending on whether you have a single or double allele (a gene from either one of your parents or both), supplementing with L-5-MTHF can be life changing. With that said, I have found that you have to incorporate L-5-MTHF supplement very slowly into the nutritional regimen, both in dosage amount (usually 1 milligram) and frequency where some sensitive individuals may need to take a low dose every other day, or even every third day with their multivitamin and minerals.

Dosage: As recommended on the bottle, taken with a meal.

Multivitamin/mineral supplement saved my life!

I was 24 years old, when I started taking a multivitamin and essential minerals supplement. That was in 1986, 30 years ago when supplements were thought of as quackery medicine. The reason why I started taking supplements was my high-stressed life as a full-time chiropractic medical student. Full days of classes and late hours of studying were so intense, plus I didn't have much money at the time, so I couldn't eat a lot fresh healthful foods. I had to drive back and forth through an hour of traffic from Santa Monica to Whittier 5 days a week; the stress from traffic was horrible—no different from today!

Due to my time constraint, making meals was almost close to impossible so I resorted to fast food and Tiger's Milk protein bars. Some of you may remember that old-time bar with the orange label and cute tiger on the front? It makes me ill thinking about it now how that was my morning ritual on the way to school, full of high fructose corn syrup, dairy products and peanuts. And I would swish it down with a large cup of Dunkin Donuts coffee with sugar and cream. My vegetable quota for the day was the slice of tomato and lettuce in my bologna sandwich, and I can't forget the half of an iceberg lettuce salad drenched in cheap Italian dressing that complemented my dinner of soggy noodles and plain pasta sauce from a bottle. Absolutely no nutrients at all, just empty calories!

Long story short, it all caught up to me; I started to feel extreme fatigue and was losing my youthful enthusiasm. My hair was falling out and I had bleeding gums, I had dry skin and was constipated, and I had no energy or motivation to exercise. Plus my ability to retain information was slipping which was totally disconcerting (mind you, I was a very high achiever who ended up graduating with a summa cum laude honors).

Concurrently as my irritating symptoms were popping up left and right, I was lucky enough to be enrolled in a year course of Clinical Nutrition, a very important aspect of chiropractic philosophy of health, to learn the benefits of food and vital nutrients as well as what may happen to the human body when you are deficient in them. As I was reading the course book about nutritional deficiencies, I remember saying to myself, "Oh no! I have scurvy, I have beriberi!" Ok, I was a bit melodramatic, but one thing that chiropractic and medical school will do to you after learning about the terrible diseases out there is turn you into a hypochondriac.

I immediately went to Mrs. Gooch's (before there was Whole Foods) and as I was looking at the hundreds of bottles in front of me trying to decide which multi to get, I kept on hearing my mother's voice saying "Buy quality, not quantity!" So I grabbed the most expensive bottle of multivitamins and started taking them twice a day. I doubled the amount it recommended because I wanted my body and mind to come back quickly.

I changed my diet and made sleep a priority and within weeks, my symptoms started to disappear, my energy and alertness was coming back and I was back to my energetic, fun self again! My hair of course took months to grow back, but I was elated when I saw the tiny baby hairs sprouting all over my head!

If I was put in front of 100 bottles of different types of supplements, herbal remedies and homeopathic formulas and was given a choice to only pick one bottle, without a doubt, I would choose a complete multi-vitamin/multi-mineral formula. I want the insurance that I get all of the essential nutrients daily. Relying on just food is not an option anymore. Studies show that when you are deficient in essential nutrients, your cells get stressed, you damage your mitochondria, your tissues weaken, mutate or die and disease settles in. Take a multi, even if you don't feel anything. It may save your life!

Did you know that scientists have identified only fifty factors that are essential to human survival, for human life? Forty-six are orthomolecular nutrients, and the rest are air, water, calories and sunlight!

46 Nutrients Essential for Human Life
13 Vitamins
21 Minerals
10 Amino Acids
2 Essential Fatty Acids

Forty-six nutrients are essential to human survival, but most mainstream doctors don't even recommend a multi-vitamin because they still believe we get plenty of nutrients in our food.

I beg to differ. The recommended daily allowance (RDA) levels are, in my opinion, way too low to support optimal health. The RDA is the level of vitamins and minerals that will *prevent* disease, not *support* health.

Who wants just enough levels of nutrients, so you don't have a nutrient deficiency disease such as scurvy or beriberi? Don't you want optimal levels of nutrients, so your mitochondria and body can function at the optimal high performance level?

I did some math and calculated that to get 1000 mg of vitamin C at one time (which I do at least twice a day), you have to ingest one medium orange, one medium papaya, and 9 cups of veggies and fruits. There is no way I can eat that in one sitting, I will surely have a case of a swollen belly and a ton of gas fast!

Also, because our soils are depleted of minerals due to poor farming practices and adulterated by chemicals such as pesticides, herbicides and fungicides, the vegetables and fruits from these farms have lower levels of nutrients in them.

Nutrients are vital for life. So, yes, the inconvenience of popping a pill once a day ensures you are supporting the basic foundation for your mighty mitos and body.

To get learn more details on vital nutrients you need to thrive optimally and prevent disease, my *Essential Nutrients: Quick Reference Guide* is available in the Resources section in the back of the book.

Glutathione (GSH)

Glutathione (GSH) is the master of all antioxidants and production of it depends on the availability of three amino acids—**glutamate, glycine and cysteine.** It will help neutralize free radicals, viruses and all types of toxins, and will also facilitate DNA synthesis and repair. Listen to my Wellness for Life radio show on all the benefits and power of glutathione, the link is provided for you in the Resources section.

I recommend to my patients a sublingual glutathione product that has an etheric liposomal delivery system where the glutathione is directly absorbed through the buccal mucosal tissue in the mouth and bypasses the gut. It has the best delivery mechanism I have experienced so far, since glutathione in pill form has absorption challenges through the gut. I personally take about 300 mg per day.

Did you know that folate (vitamin B9) boosts glutathione production? Studies have shown that many people in the U.S. don't get enough folate and recently with more data coming out regarding methylation issues and methyl genetic analysis, results show that there are 35–75 percent of the U.S. population who have genetic variants that put us as risks of low folate. If you are deficient in folate, you may be deficient in glutathione production.

Dosage: 200–500 mg per day depending delivery system

L-Carnitine

L-carnitine is the key ingredient to help shuttle lipids (fat) through the mitochondrial membrane and into the mitochondria, where it can be burned as fuel to produce energy, ATP. It's considered a nonessential amino acid since your body makes it in the liver and kidneys. It is readily stored in the skeletal muscles, heart, brain, and sperm. However, genetic variants may reduce the production of this essential nutrient.

Interestingly, I found out recently that I have two genetic variants (LOC553103 gene and SLC22A5 gene) that block my endogenous production of l-carnitine, basically wiping out my ability to produce carnitine by 50 percent. That is a huge limitation in my book. But sure enough, this was one formula I started taking about the same time as the multi formula 30 years ago! Intuitively, I knew my body really loved carnitine and wanted it daily, even throughout my pregnancy with my son. I don't know many people who have been taking this formula as long as I have,

but what I do know is that l-carnitine has been instrumental to my mitochondrial health and my energy metabolism as well as helped me maintain a healthy figure. Who doesn't want that!
Dosage: 400–500 mg per day

CoQ10

CoQ10 Enzyme or Coenzyme Q-10 (CoQ10) is a rate limiting nutrient factor in the electron transport chain and has a key role in producing ATP in the Krebs Cycle and improving energy levels in all tissues.

Coenzyme Q-10 levels are highest in the first 20 years of our lives and slowly go down, so its levels may be associated with the aging process. Other benefits include helping to regulate insulin levels for those with diabetes and lowering blood pressure.

Before I go into the next mitochondrial-enhancing nutrient, I want to discuss briefly of a unique form of CoQ10 with which I am currently running a patient study, and I believe the symptom surveys collected are indicating that it may stands out above the rest of CoQ10 formulas I have used. It's called MitoQ®.

Traditional CoQ10 supplements have difficulty getting into the mitochondria because of their complex double membrane structure, so you have to take high dosage levels of the antioxidant.

The breakthrough came when MitoQ® researchers put a positive charge on CoQ10, creating an electrostatic gradient. Because mitochondria are negatively charged (negative attracts positive), the new positively charged CoQ10 literally flooded into the organelle between **800 and 1200 times** that of regular CoQ10.

After taking the formula for a short period of time, some of my patients with chronic fatigue syndrome and neurological conditions such as Multiple Sclerosis and Parkinson's disease have started to experience more frequent energy bursts and are able to sustain longer bouts of body and brain energy throughout the day. This is very promising in the world of mitochondrial medicine.

Dosage: 100–300 mg of regular CoQ10, 10–20 mg of MitoQ® per day

Essential Amino Acids

Essential Amino Acids help regenerate mitochondria and ATP production. They are the building blocks of protein that are indispensable for human life because they cannot be synthesized within the human body; therefore, they must be supplied through the diet.

As mentioned previously, all protein is not created equally and much of the protein you eat in food is just wasted and eliminated through your feces. As we age, our assimilation of proteins decline considerably and by the time you are 80 years old, you have reduced your muscle mass by 40 percent. Mitochondria makes up 10 percent of our muscle mass, so with less muscle you end up with less energy!

To prevent age-related muscle loss and mitochondrial impairment, I highly recommend taking an essential amino acid formula that has high levels of branched-chained amino acids: leucine, isoleucine and valine. As mentioned earlier, if you are a vegetarian or vegan, amino acid supplementation is imperative to mitochondrial health!

I take Super 8 Aminos (for more information, check the Resources section in the back of the book) daily, and I believe taking essential amino acids was an integral part of my healing from sarcopenia and cervical neck pain.

Dosage: 1500–2500 mg twice a day, in between meals.

Mitochondrial Membrane Lipid: Phosphatidylcholine And Omega 3 Fatty Acids (Docosahexaenoic Acid, DHA)

In 2006, in the Journal of Chronic Fatigue Syndrome, Dr. Garth Nicholson and Dr. Rita Ellithorpe published an article titled: "Lipid Replacement and Antioxidant Nutritional Therapy for Restoring Mitochondrial Function and Reducing Fatigue in Chronic Fatigue Syndrome and other Fatiguing Illnesses"(the full article link is available in the Resources section.) Their 8-week patient study involved utilizing Lipid Replacement Therapy (LRT) (in addition to antioxidants) as a key component to restoring mitochondrial and other cellular membrane function.

Phosphatidylcholine and omega 3s (specifically DHA) were the two nutrients essential for the protection of the inner and outer mitochondrial membrane from oxidative and other damage, but also, these two supplements were intrinsically required for proper inner mitochondrial membrane function and electron transport chain via the Krebs cycle. Put simply, you need DHA and Phosphatidylcholine for optimal mitochondrial function and to repair mitochondrial membrane and ultimately to make energy.

Dosage: 1000 mg twice a day with meals

Magnesium

Magnesium has been linked to the reduction of mitochondrial DNA mutation. About 50 percent of us are magnesium deficient! Additionally, magnesium supports other health benefits such as helping maintain a healthy heart and muscles and a powerful immune system. Magnesium is also a key cofactor for energy production via the methylation pathway, particularly the methionine pathway.

Dosage: 300 mg twice a day, with or without meals (reduce levels if you experience diarrhea)

Pyrroloquinoline Quinone (PQQ)

Pyrroloquinoline quinone (PQQ) is a potent antioxidant that encourages the production of new mitochondria (mitochondrial biogenesis), and influences the size number and density of mitochondria which leads to more ATP and energy production. That's why it's different than supplements like l-carnitine, which works by shuttling fat into the mitochondria for energy production. In addition, like CoQ10, PQQ works as an antioxidant and provides powerful defenses against mitochondrial impairment.

While the body cannot make PQQ, it can be found in a variety of foods including parsley, green tea, green peppers, kiwifruit and papaya. It is an enzyme cofactor possessing antioxidative qualities.

Dosage: 10–20 mg per day in the morning, with or without a meal

Buffered Creatine Monohydrate

Buffered Creatine Monohydrate is a very stable, bioavailable form of creatine and provides extra support for energy production, muscle enhancement and athletic performance. There are unpleasant side effects from taking high doses creatine supplement such as nausea, diarrhea, cramps and bloating due to the conversion of creatine into the metabolite, creatinine. To prevent side effects, look for buffered creatine monohydrate.
Dosage: 1000–2000 mg per day

Rhodiola Rosea

Rhodiola rosea, also known as "golden root," is a popular *adaptogenic* herb, which means it works in the cells to normalize their function and stimulate healing. This herb supports the adrenal glands and can help the body deal with stress, anxiety, and fatigue. However, research[13] also shows that Rhodiola rosea is a powerful herb for enhancing mitochondrial energy production. It works by activating the synthesis of ATP in mitochondria as well as being a powerful antioxidant.

Dosage: 200–400 mg per day, in the morning

Alpha-Lipoic Acid (ALA)

Alpha-lipoic acid (ALA) has a number of interesting functions. It can help restore other antioxidants if they've been used up in the body. It also has the unique claim to fame as the only antioxidant that can be shuttled into your brain easily. It can be especially helpful for people suffering from Alzheimer's for this reason.

Dosage: 100–300 mg, divided twice a day, with a meal

NADH: Reduced Nicotinamide Adenine Dinucleotide

NADH: Reduced Nicotinamide Adenine Dinucleotide is a nutrient essential to human health, and without it the person will display symptoms such as dermatitis, diarrhea, dementia and eventually death, called *pellagra.* Studies[14] show that stabilized oral NADH can reduce symptoms of fatigue, cognitive dysfunction and dementia, and other neurological disorders such as Parkinson's disease. Niacin, (B3), nicotinamide or nicotinic acid can be ingested as precursors supplements to NADH and are also found in seafood and animal protein, avocado, nuts, green peas, sunflower seeds and chia seeds.

Dosage: 10–20 mg per day, in the morning

Astaxanthin

Astaxanthin is one promising nutrient that may improve your endurance by boosting mitochondrial antioxidant defenses. It is a pinkish orange carotenoid found only in two ocean sources: a microalgae rich in reddish pigment and seafood such as tiny crustaceans called krill and wild salmon (the darker the orange hue the better; my favorite is wild sockeye salmon).

Don't be fooled by farm-raised "pink" salmons, they are fed commercial feeds containing a synthetic version of the natural pigment astaxanthin derived from petrochemicals, as revealed by Randy Hartnell, CEO and Founder of Vital Choice Seafood. Without this artificial astaxanthin, farm raised salmon would be the color of regular fish but, more importantly, they are found to have ten times higher levels of toxins such as polychlorinated Biphenyl (PCBs) and dioxin levels compared to wild caught salmon. When it comes to neutralizing free radicals, astaxanthin can be as much as sixty-five times more powerful than vitamin C and fifty-four times stronger than beta-carotene! Now that's a super antioxidant!

Dosage: 5–10 mg per day with meals

Vitamin D

Vitamin D is an orthomolecular prohormone, and if deficient can challenge energy storage and energy production during recover phase from moderate exercise. Some individuals have genetic variants such as VDR Taq or VDR Bsm that block the metabolism and utilization of vitamin D3. I am one of those people; I have double VDR Taq genetic defects, which means I have to orally ingest Vitamin D3 for life. No matter how many hours of daily sun exposure (I am known to have a suntan all year around!), I can't get my blood levels of vitamin D 25-OH over 35 mg/dl, which is a low Vitamin D level. I take 2000 IU during the summer and 5000 IU during the winter.

Dosage: 1000 IU per day (get a vitamin D 25-10 blood test for best dosage levels)

Vitamin C

Vitamin C is a well-known antioxidant and required to synthesize l-carnitine. It takes 11 cups of fruits and vegetables to get at approximately 1000 mg of Vitamin C. High levels are found in citrus fruits, red peppers, berries, kiwifruit, avocados, beets, papaya, pineapple and sweet potatoes.

Dosage: 500 mg three times a day with meals

Vitamin E

Vitamin E is an antioxidant we're all familiar with and comes naturally in lots of foods. Look for foods rich in *gamma tocopherols and tocotrienols* such as coconut, red bell peppers, walnuts, cranberry, rice bran, ground cinnamon, flax seeds, palm oil, cocoa butter and gluten-free oats.

Dosage: 400–800 IU per day with meals

NAC (N-Acetyl Cysteine)

NAC (N-Acetyl Cysteine) is a precursor nutrient for the synthesis of glutathione. It's also known for its potent antioxidant properties. It's great to use during allergy season and respiratory infections to break up the congestion and clean up the lungs as a mucolytic. I call it the all-natural decongestant.

Dosage: 300–600 mg per day with or without meals

Melatonin

Melatonin is an antioxidant and hormone produced by the pineal gland to promote deep uninterrupted sleep and other anti-aging properties. Melatonin is a cofactor for the mitochondria and stimulates directly Complex 1 and Complex 4 of the electron transport chain in the Krebs Cycle. Studies show that there are health benefits with oral melatonin supplementation and Type 2 Diabetes, hypertension, migraines, fibromyalgia and more. Some people are very sensitive to melatonin (can get tired or wired) and can take only .3 mg per night. Never take melatonin during the day; it can throw off your natural biorhythms. We all need our sleep! I usually recommend my patients to take the melatonin 30 minutes before bedtime and start slowly, as low as .3 mg to as high as 4 mg per night. Read more on melatonin in Chapter 15: Healthy Lifestyles for Mitochondrial Health.

Dosage: .3–1mg 30 minutes before bedtime

D-Ribose

D-Ribose is a five-carbon monosaccharide used by all living cells for optimal cellular energy production and nucleotide synthesis (which are required by heart and muscles cells in order to make ATP). It can help boost muscle strength, promote cardiovascular health and reduce the symptoms of chronic fatigue syndrome.

Dosage: 2 grams per day (before or after exercise)

Shilajit

Shilajit is a resin exudate that oozes from the rocks of the Himalayas during the summer months with anti-aging and life enhancement properties. It contains a substance called fulvic acid that provides potent antioxidant properties and magnifies the delivery of nutrients inside the cells to recharge the mitochondria.

Shilajit also enhances the oxygen carrying capacity, improving blood circulation and combat high altitude sickness. I personally use a powdered form of nutritious superfoods and herbs, that includes both shilajit and ho shou wu, as part of my tonic called Jing Jing (for more information, check Resources section in the back of the book.)

Dosage: ½ teaspoon of shilajit powder in the morning

CHAPTER 14

Mighty Mito Nutrition Plan Step 5

Oxygenating Pudding, A Mitochondrial Superfood

OXYGENATING PUDDING IS PERHAPS THE single best food you can eat for mitochondrial health since it is so easily absorbed and converted to powerful ATP. By eating more of it, you can start feeling more boundless energy, sharpen your brain clarity, and have a powerful vibrant body!

Where Did Oxygenating Pudding Come From?

This pudding creation of mine came about after studying in detail Dr. Johanna Budwig's brilliant work. Dr. Budwig was a biochemist and physicist in the 50s and was the first to identify that there were key health differences between saturated and unsaturated fats. One of her mentors was Dr. Otto Warburg. As mentioned in chapter 3, he won the Nobel Prize for Physiology in 1931 for discovering the mechanism for mitochondrial cell respiration. Dr. Warburg noted that cancer cells mainly generate energy by fermentation, the anaerobic (non-oxidative) breakdown of glucose inside the cell, *not* in the mitochondria. He also discovered that cancer cells cannot survive in the presence of high levels of oxygen, as found in an *alkaline* state. You'll soon see how his insights tie in.

In 1952, Dr. Budwig realized that the highly unsaturated omega 3 fats (such as those found in flax oil) were the undiscovered decisive factor in respiratory enzyme function. Better respiration, more oxygen for the mitochondria. Highly unsaturated fats, such as Omega 3 fatty acids are critical to normal cell metabolism and without adequate supplies, cells begin to produce energy from glucose through fermentation. This is the connection to Dr. Warburg's work: that cancer can thrive where there is oxygen deficiency, and the acidic (non-alkaline) state in the human body that it creates.

Increase oxygenation to increase life!

Dr. Budwig found that not only did the Omega 3s in flax oil help prevent cancer, it prevented and reversed heart disease, arthritis, and other inflammatory diseases. Therefore, what we want is an alkaline, highly oxygenated environment in our bodies (with Omega 3s to help the respiratory process) so that our mitochondria can produce energy in an aerobic process. We should decrease anything that makes our body more acidic or relies on fermentation, which is how cancer cells survive. This is our best bet for preventing or reversing all kinds of disease.

Dr. Budwig wanted to find a superfood to increase the oxygenation of the human tissues and as she experimented further, she created the *oil-protein diet,* which combines flax oil, with high-sulfurated proteins found in Quark, a European dairy product similar to cottage cheese.

She discovered that by mixing these two specific foods together, the flax oil and high sulphureted protein found in quark became water soluble and electron rich. It was, in essence, almost the perfect food for our digestive system and absorption by mitochondria. These electron rich molecules jumpstart normal cell biology, and shift energy back to the mitochondria and increase our oxygen utilization.

Dairy-Free/Paleo Oxygenating Pudding

As you've just learned, our mitochondrial healthy nutrition program eliminates dairy because of the sensitivities most people have to it. So I put on my science geek hat and started digging deeper, researching and trying all different types of non-dairy proteins to substitute for the Quark. After trying all types of proteins, from rice protein and hemp protein to soy protein and beef protein and even mung bean powder, I finally hit on the answer: I needed a highly sulfurated protein (high levels of methionine and cysteine, for other science geeks out there), and the answer was *egg white protein*. It was truly a breakthrough moment for me!

So here's what you need to make this tasty mitochondrial superfood. And please note, do not try this recipe if you have an egg allergy or sensitivity.

1. High lignan flax oil by Barleans, which is a company with high standards of manufacturing. Why high lignan? Because the phytonutrients in flax seed are known to possess impressive cancer-preventive properties.

2. Egg white protein. I like both vanilla and chocolate flavor. Use hormone-free and antibiotic-free egg white protein with no artificial flavors or sweeteners. I use one that is sweetened by stevia.

3. Next, you can choose a non-dairy, unsweetened milk alternative. I use almond milk, but coconut is fine too. You can also use unsweetened coconut yogurt to get a thicker pudding consistency.

4. Get an immersion blender with a low setting. Don't use any other kind of blender or mixer because blending at high levels will heat up the flax oil and that can denature the Omega 3s.

5. Get a glass cup or mug that is narrow enough to submerge a mixing stick but not so wide that it can't blend all of the ingredients well.

Then you'll want to mix the following amounts of the ingredients together:

- Add 4 tablespoons of unsweetened non-dairy milk or non-dairy yogurt
- Add one scoop of the egg white protein
- Add 2 tablespoons of flax oil after shaking the bottle of oil

There is always a two-to-one ratio between the milk or yogurt and the flax oil. Do not change this ratio in any way because I have discovered this is the exact amount that works best with one scoop of the egg white protein: twenty-three grams of egg white protein. Any more or less can affect the formula and amount of oxygenating properties.

Next, blend with the low setting, and you want to pulse it on and off, slowly blending without heating it up. Move the blender up and down. You will notice that it starts to emulsify nicely but make sure there are no visible oil droplets. Just blend it for 1 minute.

Voila! There is your oxygenating pudding. You'll want to eat it in the next 30 minutes while the oil is still stable. You can even add blueberries, raspberries, or flaxseed powder to make it tastier; add it after you have made the pudding mixture. Just do make sure you grind up the flaxseed every time you make the pudding, flax powder from health stores can go rancid very quickly.

As long as you are not allergic to the ingredients of the oxygenating pudding, you can eat it on a daily basis. I personally have the oxygenating pudding once a day, as a midmorning or mid-afternoon snack. I must tell you the other day, I added 1/2 teaspoon of organic decaffeinated freeze dried instant coffee to the last 10 seconds of the blending, and it tasted just like coffee ice cream, melted of course!

If you find yourself having loose stool after trying the Oxygenating Pudding, it may be a sign that you are having trouble digesting fats and oils. If so, make the pudding with a smaller amount of flax oil, such as 1 teaspoon and build up to 1 tablespoon and then up to 2 tablespoons per recipe to get the full oxygenating benefit.

If you feel fine eating 1 tablespoon, then slowly work yourself up to 2 tablespoons. One more thing, taking a digestive enzyme supplement with ox bile and lipase as part of the formula may also help you digest fat better.

CHAPTER 15

Healthy Lifestyles for Mitochondrial Health

THE FINAL STEP TOWARD ULTIMATE mitochondrial health is to improve imbalances in our psychological, emotional and physical state. These guidelines, while applicable to most people, should never override proper psychological and/or medical care, especially for those suffering from depression, anxiety or other psychological issues. They are, however, often some of the very keys in dealing with even the hardest moments and psychological conditions, since they all go to protect and produce healthy mitochondria, which is a key for emotional wellbeing.

At the most basic level, we know that stress being a normal part of life, releases cortisol, the stress hormone, which can slow down our metabolism and in the long term, eat away at muscle. Because we experience stress as "fight or flight," our bodies are preparing us for what it imagines are "real" threats in the world. Cortisol was helpful to release sugars in our bloodstream when we need to run away from lions on the savannah. It's still helpful when we need to jump out of the way of a speeding car. It's not as helpful when we're simply fretting over how much the car repairs are going to cost for 3 days on end. Stress, in our modern society, builds up slowly over time, meaning more and more cortisol is released and stays in the bloodstream. It's not a "one-time solution" to an immediate danger. We experience events in everyday life as a sort of "threat" and produce cortisol to compensate. This can be *extremely* harmful to our health.

Because cortisol was used for emergencies in our evolutionary heritage, it also works to slow down our digestion. Again, we needed energy to run away from predators. But what that means for us nowadays is that we're not extracting energy from our food by living every day with stress and that means less ATP coursing through our system.

So shouldn't we do everything we can to accept the stress we *do* have and not add more stress by fighting it? But shouldn't we also try to learn how to not feed in to creating more of it? Shouldn't we learn to be kinder when we are suffering, but also creating the healthy emotional and physical habits that will reduce its overall impact on our lives?

This is the first step. If right now, you feel stressed, there's no reason to feel guilty, shamed, or "stressed" about stress. We all feel it and are all imperfect. It's an illusion that anyone is happy all the time and there's no reason to expect this of yourself. See if you can treat your own stress as you would the stress of a loved one or friend. Give yourself the compassion that you would give them. Acceptance is the best way of dealing with our emotional difficulties.

From there, you have a good foundation to incorporate some lifestyle changes that will reduce stress in your life. And that will only lead toward better mitochondrial *and* emotional health!

Breath and the Power of Oxygenation

Since you now know just how important oxygenation is for your mitochondria, you probably won't be surprised to realize how important breathing properly is for those little organelles. But have you ever stopped to think about your breathing and if you're doing it in the best way for your body?

The truth is we can drastically change our habitual breathing pattern for the better. Most of us, especially in our stressed-out, faced-paced society tend to be chest breathers, meaning that we take shallow breaths without really using our abdomens. Breathing from the chest, you tend to get less oxygen and therefore less power for your mitochondria. Additionally, your lungs lose anywhere between 9 and 25 percent of their capacity with each passing decade, starting in your mid-20s.

We want to change that by changing your breathing.

The trick is to begin breathing from your belly, which not only has the benefit of getting you more oxygen, but it also makes you feel calm and content. Belly breathing allows our body to move oxygen down to areas within our lungs where most of our blood circulation occurs. Here's a simple practice that you can incorporate starting today, and that with daily practice, sometimes in as little as 2 weeks, will become natural for you. By beginning a belly breathing practice, you will naturally start to take more deep breaths when you're not even thinking about it.

The best way to do this is to sit upright and place one hand over your chest and one hand over your belly. Make sure you inhale and exhale through your nose. Keep your mouth closed. The chest can extend a little bit, but you want the majority of your breathing to come from your abdominal region. That is, the hand on your belly should rise much more than the hand on your chest. If you don't get it at first, don't worry! We've been programmed to breathe from our chest, and some of us for 30 or 40 years of our lives.

As you inhale through your nose with your mouth closed, count for four seconds, this is usually far more time than we usually inhale during our day-to-day stresses. Pretend you're filling up a balloon that is within your belly. Don't strain, just take in what feels like a comfortable amount of air. As you reach four seconds, exhale through your nose with your mouth closed and count to six seconds. Keep a four-to-six second inhale-to-exhale ratio, just focusing on the breath. I call this my 4:6 Breathing Rule. And you may notice your body relaxing, the tension releasing from your face, nervousness lifting, more energized and a warm sense of well being!

It's best to practice this for about 10 minutes at a time, three times a day. If putting a hand on your chest and belly feels a little silly while you're at work or in public, you can simply try to pay attention to how much your belly rises versus how much your chest rises. As you begin your practice, you might put a timer on your cell phone so that you practice it in the morning and in the evening. It is, quite simply, one of the most relaxing and oxygenating things you can do for your body.

More oxygen plus less stress equals mightier mitos!

Laughing for Life Exercise

Did you know that an average 4-year-old child laughs 300 times a day, and the average 40-year-old adult, only four times?

Researchers have shown the following benefits of laughing out loud:

- Reduces levels of stress hormones such as cortisol, epinephrine, and norepinephrine
- Increases health-enhancing chemicals such as endorphins and our cells love it
- Help secrete feel-good neurotransmitters such as serotonin
- Decreases pain and inflammatory chemicals
- Boosts infection-fighting antibodies
- Improves blood flow to the heart, lungs and brain.

These physical benefits also help:

- Our mood and behavior
- Improve our positive outlook on life
- Us to be friendlier and more engaging
- Us feel more attractive and more alive!

In addition, the physical action of laughter improves your breath capacity and exercises the diaphragm, increasing oxygen delivery to every cell in your body to power up your mitochondria.

Medical scientists have acknowledged that laughter therapy can help improve the quality of life for anyone suffering from chronic illnesses.

I learned the following laughing exercise from one of my mentors Mary Morrissey, but it was originally developed by Reverend Masaharu Taniguchi, the founder of the Japanese movement called *Seicho No Ie* (Truth of Life).

Reverend Taniguchi believed that you could be happy and then laugh or laugh to be happy. He also believed that you could learn to laugh even if there isn't anything particular funny to laugh about.

He asks you not to laugh like "Ho ho ho" or with a "Hee hee hee." The correct way to laugh is with a "Ha haha haha haha"

But first he suggests you recite his beautiful ***Laughter Poem*** out loud:

Spirit of Joy
In dwelling me
Ever present
I call you forth—
Now!

Next, while you are seated or standing, put both hands on your belly and tip your head back and start laughing out loud, bring your head forward and repeat!

Ha haha haha haha!

Do it a couple of times, even if there is nothing to laugh about. Just do it and see what happens. You will naturally start laughing just from the joy and the fun of it! Laughing is very infectious, and if you do it with a friend or group of people, wow there is so much laughter power generated that your body and mighty mitos will be sparking!

To watch a quick video where I am showing Chris Smith of The Campfire Effect on how to perform the Laughing for Life exercise, go to the Resources section for the link.

Exercise and Increasing Stress Tolerance

We have already covered the importance of burst training exercises in culling damaged mitochondria and increasing oxygenation. In addition, choose an exercise that also focuses on breathe work, such as yoga, tai chi, or even some martial arts, and you have a super recipe for mitochondrial health. But let's not forget the role exercise can have in stress reduction.

When you exercise, you release endorphins, which make you feel great. Basketball, hiking and tennis, can all take your mind off your stress and show you just how fun and enjoyable life can be. You aren't *ignoring* your stress, but you're using your body to enjoy life *despite* your stress. Exercise is one of the best "flow" experiences, in which we become so immersed in the moment. Those flow moments are a key for reducing stress and living more fully. At the physiological level, exercise also raises the threshold at which cortisol may be released, meaning that it will take higher levels of "stress" for you to become stressed, and therefore will end up reducing the overall release of cortisol. That's why when we've been exercising regularly, we often feel that the "little things" don't get to us in the same way that they did before we were exercising. There's a new biological threshold we've reached for when we become stressed.

Sleep and Mighty Melatonin

There's one essential thing that most Americans are deprived of, and that is *sleep*. Many people function, or think they function on 5 or 6 hours of sleep a night. And while there is real genetic and biological diversity in how much sleep we actually need, the truth is most people need between 7 and 9 hours of sleep.

What's the Mitochondrial Connection?

As you're winding down and then asleep, your body begins to produce increasing levels of melatonin. Melatonin, as mentioned earlier in the supplement section, is a key antioxidant and helps with the electron transport chain in the Krebs cycle and ATP generation in the mitochondria. So sleep is literally empowering your mitochondria to produce energy! Melatonin is also "selectively taken up by mitochondrial membranes, a function not shared by other antioxidants," meaning that it is one of the only antioxidants that really focuses on reducing oxidative stress on the *mitochondria themselves,* rather than all parts of the cell.[15] It's a healing chemical for your mighty mitos. So why deprive yourself?

There's also a good reason why it's ideal to follow a sleep pattern that mimics the night-day cycle: melatonin levels in our bodies generally tend to rise at night because our bodies are *asking for sleep* as part of their natural biorhythms. This is why so many people who work the graveyard shift, even though they adapt over time, never feel quite as good as when they go back to a more natural sleeping pattern.

How to Improve Your Sleep

It's not just working a graveyard shift that disrupts our sleep cycle. Too much light during late hours, tapping away on tablets (especially in bed!), watching TV, or stimulating activities, like working late into the evening, can give our brain the signal that it's not time to sleep, but rather time to be active. Restlessness, stress, and even insomnia are the natural results.

To improve your sleep cycle and draw on the power of melatonin for your mitochondria, the best thing you can do is to slowly close down all the day's activities at least an hour *before* you go to bed. This also means dimming or shutting off your lights. Instead of taking your tablet to bed, bring a book with you and use a small book light, as studies have shown reading before bed can help you sleep better. It's also especially important to avoid alcohol altogether and caffeine in the 6 to 7 hours before bed, as these chemicals can be great sleep disruptors. Additionally, it's best not to use your bed for anything other than sleep and being intimate with your partner. That way, your brain naturally associates the bed with sleep. If you suffer from insomnia, it's often best not to lie in bed for more than 15 or 20 minutes, as we often start ruminating, but rather to get up and do something else and return to bed the next time you feel sleepy.

By improving your breathing, exercise, and sleep you will go a long way toward reducing your stress, improving oxygenation, and draw on the power of melatonin for mitochondrial function. It all begins with accepting the stress you have, knowing we are all imperfect creatures, and from there, trying to make small but lasting changes in how we relate to our breath, body, and others around us.

CHAPTER 16

Conclusion

YOU'VE NOW LEARNED A POWERFUL formula for increasing your mitochondrial health at the deep, biological level. This formula can help you not only feel more energy and emotional well-being, but hold off or even reverse some of the most prevalent diseases of our time: diabetes, high blood pressure, heart disease, dementia, Alzheimer's disease, fibromyalgia, pain syndromes, autoimmune diseases, and even cancer.

We know just how powerful the mitochondria are because they are the power plants of your body, the generators that make everything run. We began by examining how to avoid free radical damage and oxidative stress by reducing our exposure to environmental toxins like polluted air, unhealthy water, UV radiation, and volatile organic compounds. From there, we looked at the power of exercise in culling and getting rid of damaged parts of the mitochondria. This left only the healthiest parts of the mitochondria so they could fuse with other, strong mitochondria. Finally, we looked at several ways to produce new healthy mitochondria and protect the ones we have, including changing our diet, using powerful supplements, and reducing stress through proper breathing, laughter, exercise, and sleep.

Implementing these changes won't always be easy, or come naturally. But as my own story shows, recharging your mitochondria is one of the most healing things you can do to heal and prevent disease. I went from a traumatic brain and neck injury that threatened my health and career, to feeling boundless energy, experiencing more brain clarity, and sustaining a powerful body, once I uncovered the secrets of the mighty mito. That's why I know that no matter how bad you're feeling now, you can begin to implement some of these life-changing guidelines today. And the sooner your journey begins of empowering your mitochondria to work for you, the sooner you'll begin to feel the vibrancy, joy, and health that our bodies are capable of.

To your vibrant health and ultimate wellness,

Dr. Susanne

Dr. Susanne

Appendices

Ultimate Wellness Food Checklist

Do Eat These Foods

Protein
- All game meats
- Beef
- Chicken
- Eggs
- Fish/seafood (1x/week only)
- Pork
- Turkey

Fruits
- Avocado (limit ¼)
- Banana (small)
- Blueberries
- Boysenberries
- Cantaloupe
- Cranberries
- Coconut
- Dragon fruit
- Durian
- Kiwifruit
- Lemons
- Limes
- Mandarin oranges
- Melons
- Oranges
- Papayas
- Passion fruit
- Pineapple
- Raspberries
- Rhubarb
- Star anise
- Star fruit
- Strawberries
- Tangelos

Vegetables

- Arugula
- Bamboo shoots
- Bean sprouts
- Beets (limit to 4 slices)
- Bok choy
- Broccoli (limit ½ cup)
- Brussels sprouts (limit ½ cup)
- Butternut squash (limit <¼ cup)
- Carrots
- Celeriac
- Celery
- Chives
- Cucumber
- Eggplant
- Endive
- Ginger
- Green beans
- Kale (cooked)
- Lettuce
- Olives
- Parsnips
- Peas (limit <¼ cup)
- Potatoes
- Radish
- Red bell pepper
- Rutabaga
- Scallions (green portion)
- Spinach
- Summer squash
- Sweet potatoes (limit ½ cup)
- Taro
- Tomatoes
- Turnips
- Water chestnut
- Yams
- Zucchini

Fresh Herbs

- Basil
- Chili
- Cilantro
- Coriander
- Ginger
- Lemongrass
- Marjoram
- Mint
- Oregano
- Parsley
- Rosemary
- Thyme

Nuts/Seeds (Limit one handful)

- Almonds
- Chia seeds
- Flax seeds
- Macadamias
- Pecans
- Pine nuts
- Pumpkin seeds
- Sesame seeds
- Sunflower seeds
- Walnuts

Gluten-Free Grains

- Amaranth
- Arrowroot
- Brown rice
- Gluten-free oats
- Gluten-free bread/cereal products
- Millet
- Polenta (corn)
- Psyllium
- Quinoa
- Sorghum
- Tapioca

Milk Alternatives

- Almond milk
- Coconut milk
- Gluten-free oat milk
- Hemp milk
- Rice milk

Sweeteners

- Erythritol
- Stevia

All Healthy Oils and Butters

- Avocado oil
- Chia seed butter
- Coconut oil/butter
- Flaxseed oil/butter
- Ghee (second choice-butter)
- Grapeseed oil
- Hemp seed butter
- Macadamia nut oil
- Olive oil
- Sesame oil
- Walnut oil

Do Not Eat These Foods

Fruits

- Apples
- Apricots
- Avocadoes (>¼)
- Blackberries
- Canned fruit in natural juice
- Cherries
- Concentrated fruit sources
- Dried fruit
- Fruit juice
- Mangoes
- Mulberries
- Longon
- Lychee
- Nectarines
- Peaches
- Pears/Asian pears
- Persimmons
- Plums
- Prunes
- Watermelon

Vegetables

- Artichokes
- Asparagus
- Beetroot
- Broccoli (>½ cup)
- Brussels sprouts (>½ cup)
- Button mushrooms
- Cabbage
- Cauliflower
- Chicory
- Fennel
- Garlic
- Green bell peppers
- Leeks
- Mushrooms
- Okra
- Onions
- Shallots
- Snow peas
- Spring onion
- Sweet corn

Legumes

- Baked beans
- Chickpeas
- Kidney beans
- Lentils
- Soy beans (tofu, tempeh)

Gluten Grains

- Barley
- Couscous
- Kamut
- Rye

- Spelt
- Tricate
- Wheat

Miscellaneous

- Alcohol (beer, wine)
- Cashews
- Dandelions
- Inulin
- Peanuts
- Pistachios

All Milk Products

- Buffalo milk
- Cheese (all types)
- Cow milk
- Cottage and cream cheese
- Custard
- Goat milk
- Ice cream
- Mascarpone
- Ricotta
- Sheep milk
- Yogurt

Sweeteners/Sugars

- Agave
- Fructose
- High fructose corn syrup
- Honey
- Isomalt
- Maltitol
- Mannitol
- Sorbitol
- Sucrose/cane/beet sugar
- Xylitol

Mighty Mito Recipes

The majority of the following recipes are from my own personal meal plan that I prepare and enjoy with my family and friends. They are delicious, savory and filling. I don't count calories; I just want nutrient-dense foods that are tasty and ready to give me power and endurance. Most meals I create are broken down into veggies, proteins and complex carbohydrates. I usually sprinkle a small amount of extra healthy oils (olive, avocado, and sesame) right before serving; remember fats are mitochondria's allies!

The ingredients found in each recipe are from my Do Eat list (Ultimate Wellness Food Checklist). These recipes are to give you a basic idea of how you can change up any of your favorite recipes or from your cookbooks to fit into your Mighty Mito Nutrition Plan.

If you are just starting your first 6 weeks of the Mighty Mito Nutrition Plan, or you are sensitive or allergic to any ingredients in these recipes or you see that one of the (optional) ingredients may include one of the high fermentable foods such as fresh garlic or onion, you can replace those ingredients with other herbs and spices, vegetables, or dried garlic or onion powder which are less in the fermentable carbohydrates. Just use your taste buds to find the best flavor!

Use organic, grass-fed, grass-finished, cage-free, free-range,

GMO-free and antibiotic-, chemical-, preservative-, pesticide- and fertilizer-free foods as much as possible to ensure purity and toxic-free meals. These guidelines are what it takes to up level your mighty mitos for high performance.

As mentioned earlier in Chapter 11, after 6 weeks on the Mighty Mito Nutrition Plan, start to experiment and change up your recipes by adding one new food back into your diet, every four days. Don't forget to add it to your food diary, so you can track your improvements or possible new symptoms.

I make some of these Mighty Mito recipes once a week because of how much we love to enjoy yummy healthful meals. I hope you enjoy these recipes and find that they will power up your mitochondria and restore your vibrancy and ultimate wellness you absolutely deserve!

One more thing, I would love to try some of your own Mighty Mito Recipes, so if you have any recipes you want to share, please do so by sending them to drmightymito@gmail.com. I love clean food, and cooking on the weekends for my family is one of my favorite pastimes!

Sections:

1. Red Meat
2. Poultry
3. Vegetarian
4. Breakfast
5. Soup/salad
6. Slow Cooking

Red Meat

Asian Fusion Meatballs

Preparation Time: 15 minutes

Cook time: 25 minutes

Servings: 4

3 tablespoons tamari (gluten-free soy sauce)
2 to 3 drops fish sauce (optional)
¼ cup sliced green onions (optional)
1 teaspoon minced or grated garlic
½ teaspoon fresh ginger, minced
½ teaspoon sea salt
½ teaspoon black pepper
1 pound ground beef, pork or turkey
1 tablespoon white or black roasted sesame seeds, for garnish
¼ cup chopped cilantro, for garnish

1. Preheat the oven to 425 degrees F.

2. In a mixing bowl, combine the tamari, fish sauce (optional), green onions (optional), garlic, ginger, salt and pepper. Add the meat to the bowl and mix to thoroughly combine with the seasonings. Form the meat into 16, 1-ounce meatballs.

3. Bake for 25 minutes on a rimmed baking sheet lined with parchment paper. Remove from the oven and garnish with sesame seeds and cilantro before serving.

Rosemary and Garlic Roast Lamb

Preparation Time: 20 minutes

Cook Time: 15–25 minutes plus 10 minutes rest time

Servings: 4–6

2 pounds boneless lamb loin
Sea salt and pepper
3 tablespoons olive oil, divided
1 tablespoon fresh rosemary
1 teaspoon garlic powder
Sea salt to taste
Black ground pepper to taste

1. Preheat oven to 375 degrees F.
2. Pat the lamb dry and season with salt and pepper.
3. Heat 1 tablespoon olive oil in a skillet over medium-high heat until shimmering. Add the lamb loins and sear for 45 to 60 seconds on each side. Transfer to a roasting pan and let cool for 5 to 10 minutes.
4. Meanwhile, combine the remaining olive oil, rosemary and garlic powder in a small bowl. Rub the lamb generously with the mixture.
5. Roast in oven for 15 to 25 minutes, until lamb reaches desired temperature (145 degrees F for medium rare, 160 degrees F for medium, 185 degrees F for medium-well).
6. Transfer lamb to a cutting board and tent loosely with parchment paper. Let rest a minimum of 10 minutes before carving into ½-inch thick slices.

Paleo-Style Jalapeño Bison Burger

Preparation Time: 30 minutes

Cooking time: 10 minutes

Servings: 4

Burger patty ingredients:

4 small heads iceberg or baby romaine lettuce

2 pounds ground bison

3 jalapeño peppers, seeded and finely diced

4 ounces Daiyaâ Jalapeño Havarti Style cheese (optional)

2 teaspoons Celtic sea salt

1 teaspoon freshly ground black pepper

1 tablespoon avocado oil

Non-Dairy Spicy Ranch Dressing:

1½ cups unsweetened plain coconut yogurt

¼ cup plain unsweetened coconut milk

1 teaspoon horseradish

1 tablespoon fresh chives, chopped

1 tablespoon white wine vinegar

2 teaspoons hot sauce

2 teaspoons dried parsley

½ teaspoon dried dill

½ teaspoon garlic powder

¼ teaspoon onion powder

1 teaspoon freshly ground black pepper

1 teaspoon sea salt

Assembly:

	Mayonnaise or Vegetarian Mayonnaise, for serving
4	slices tomato
4	slices onion (optional)
8	bread-and-butter pickles

1. For the burger: Peel away the outer layers of each lettuce head to get a diameter of 5 to 6 inches. On opposite sides, slice two 1-inch slices from the sides of the head to form the top and bottom of the bun. Repeat with the remaining heads. Set the slices aside. Refrigerate the remainder of the lettuce for another use.

2. Mix the bison, jalapeno, Daiya cheese, salt and pepper in a medium bowl.

3. Divide into 4 patties, about 8 ounces each. Flatten each patty so that it's flat and thin, about 4½ to 5 inches in diameter and about 1 inch thick.

4. Preheat a skillet or cast-iron pan over high heat. Add the avocado oil, swirling to coat the pan. When the oil begins to shimmer, turn down the heat to medium and add the bison burgers and cook until browned on the first side, about 3–4 minutes; flip and cook for 3–4 minutes more for medium well doneness. (**Note:** Bison cooks faster and is leaner than regular beef; cooking it for only 3 minutes each side to medium will keep it moist and tender.) Let rest for 4 to 5 minutes.

5. For the non-dairy spicy ranch dressing, combine the coconut yogurt, coconut milk, horseradish, chives, vinegar, hot sauce, parsley, dill, garlic powder, onion powder, pepper and salt in a medium bowl. Whisk until smooth.

6. To assemble, take one half (the flattest half) of the iceberg slices and use it for the bottom buns. Slather each with 2 teaspoons mayonnaise and top with a bison burger patty, a slice of tomato, a slice of onion, and 2 bread-and-butter pickles. Follow that with a dollop (1 tablespoon) of the non-dairy spicy ranch dressing and then the tops of the lettuce bun. Secure with skewers and serve immediately.

Poultry

Turkey Sausage Meatloaf

Preparation Time: 25 minutes

Cook time: 40–50 minutes

Servings: 4–6

Sauce topping:

4 ounces tomato paste

4 ounces water

2 tablespoons red bell pepper, finely diced

2 teaspoons basil, chopped

2 pinches of Celtic sea salt

Black ground pepper, to taste

Meatloaf:

1 tablespoon avocado oil or coconut oil

½ small onion, finely diced (optional)

2 garlic cloves, pressed or grated (½ tablespoon) (optional)

2 eggs, beaten

1 pound ground turkey (white or dark meat)

1 pound turkey sausage

1 teaspoon oregano

1 teaspoon dried basil

½ teaspoon Celtic sea salt

¼ teaspoon black pepper—or more to taste

2 carrots, grated

1 red bell pepper, grated or diced

¼ cup fresh basil

1. Preheat oven to 375 degrees F.

2. In a small sauce pan over medium-low heat, combine the sauce topping ingredients and simmer for about 5–10 minutes while you prepare the rest of the ingredients into the meatloaf, stirring occasionally until thickens like ketchup.

3. In a medium skillet, add oil and sauté the onions (if using) over medium low heat for 2 minutes then add the minced garlic and stir for about a minute.

4. In a large-sized mixing bowl, whisk the eggs and then add ground turkey and sausage with the beaten eggs. In a small mixing bowl, combine the spices, carrots, red bell pepper, and basil. Then add the spice and vegetable mixture to the meat and egg until mixed well.

5. Line large meatloaf tin with parchment paper and fill it up with meatloaf mix.

6. Spoon about ¼ cup of the sauce onto the meatloaf, spreading it evenly. Leave extra sauce to be used for dipping.

7. Bake uncovered for 40–50 minutes or until the internal temperature of the loaves reaches 160 degrees F.

Oregano and Lemon Chicken Breast

Preparation Time: 10–15 minutes

Cook Time: 10 minutes

Servings: 4

1 pound boneless, skinless chicken breast
2 tablespoons lemon juice
1 tablespoon avocado oil
1½ teaspoon dried oregano
½ teaspoon sea salt
½ teaspoon black pepper
2 tablespoons coconut oil or ghee
Extra-virgin olive oil to be drizzled onto chicken before serving

Preparation:

1. Preheat a skillet or grill to medium heat.

2. Place a chicken breast on a cutting board and carefully slide the knife through the center so that the thickness is cut in half. Continue to slice almost completely through the chicken breast, leaving it connected in the center so that it flattens out to a butterfly shape. The chicken should now be ¼–½ inch thick at most. Repeat with the remaining chicken breasts.

3. In a large bowl, combine the lemon juice and olive oil with the oregano, salt and pepper. Add the chicken and evenly coat; allow to marinate for at least 15 minutes.

4. Brush the hot grill pan with the coconut oil or ghee, then cook the chicken for 4 to 5 minutes per side, depending on the thickness of the chicken. Fully cook the chicken.

5. When you take the chicken off the grill, drizzle with the extra-virgin olive oil.

Rosemary Chicken Thighs

Preparation Time: 15 minutes

Cook Time: 25–35 minutes

Servings: 6

Ingredients:

6 chicken thighs (bone-in, skin-on)

3 tablespoons olive oil or avocado oil

2 tablespoon balsamic vinegar

1 teaspoon rosemary

1 teaspoon black pepper

1 teaspoon erythritol

Celtic sea salt (to taste)

1. Pre-heat the oven to 375 degrees F.
2. Combine olive oil or avocado oil with balsamic vinegar, rosemary, black pepper, erythritol and salt in a small bowl to create a sauce for the chicken.
3. Drench each chicken thigh in the sauce, spreading it evenly onto the skin and bottom side.
4. Once the skillet is heated, place the thighs skin side down for 2–4 minutes, until golden brown on the skin side.
5. Flip the thighs over for about 2 minutes in the skillet, then transfer to a cookie sheet lined with parchment paper and place into the oven for approximately 20 minutes.
6. Remove the chicken from the oven and allow to cool.

Cilantro-Ginger Chicken

Preparation Time: 5 minutes

Cook Time: 30-35 minutes

Servings: 3-4

1 tablespoon ghee or coconut oil
Sea salt and black pepper to taste
6 chicken thighs with bone and skin (cut off all extra fat)
1 small onion, finely sliced (optional)
2 garlic cloves, minced or grated (optional)
½ teaspoon ginger powder or fresh ginger, minced
2 teaspoons roasted sesame seeds
¼ cup tamari
½ cup cilantro
1 teaspoon erythritol

1. Preheat the oven to 425 degrees F.

2. In an oven-safe cast iron or stainless steel skillet, melt the ghee or coconut oil, then season both sides of the chicken with sea salt and black pepper, and place skin side down into the pan for 5–6 minutes or until the skin is brown.

3. While the chicken cooks, combine the onion, garlic, ginger, sesame seeds, tamari, and more sea salt and black pepper in a small mixing bowl.

4. Flip the chicken thighs over so that they are now skin side up in the pan, then pour the sauce mixture over the chicken evenly and place the pan into the oven for 30 minutes or until the internal temperature of the chicken reaches 165 degrees F.

Vegetarian

Amaranth Quinoa Salad

Preparation Time: 10 minutes

Servings: 4

Salad:

½ cup amaranth, cooked
1½ cups quinoa, cooked
2 Roma tomatoes, diced
1 cucumber, diced
1 jalapeño, seeded and diced
½ bunch cilantro, minced
½ avocado, diced
1 tablespoon chopped chives

Dressing:

½ teaspoon garlic powder
2 tablespoons apple cider vinegar
1 teaspoon whole grain mustard
1 teaspoon mayonnaise
6 tablespoons extra virgin olive oil
Salt and pepper, to taste

1. To make salad, combine cooked amaranth, cooked quinoa, diced tomatoes, cucumber, chives, jalapeño and cilantro in a large bowl.

2. In a small bowl, whisk together garlic powder, apple cider vinegar, whole grain mustard, and mayonnaise. Then drizzle in extra virgin olive oil while whisking vigorously.

3. Drizzle dressing over salad, and garnish with avocado.

Roasted Butternut Squash and Parsnips

Preparation Time: 20 minutes

Cook Time: 45 minutes

Servings: 4

1 butternut squash, peeled and chopped
3 medium parsnips, peeled and chopped
1 teaspoon ground cinnamon
¼ teaspoon allspice
¼ teaspoon nutmeg
2 tablespoons olive oil
Salt and pepper to taste
1 cup pecans, coarsely chopped
Fresh parsley, to garnish

1. Preheat your oven to 400 F.
2. In a large bowl, combine the butternut squash, parsnips, cinnamon, allspice, nutmeg, and olive oil with salt and pepper to taste.
3. Toss the squash and turnips until well-coated with the oil and spices.
4. Spread the vegetables on a parchment lined baking sheet, and place in the oven. Bake in the preheated oven for 25 to 30 minutes.
5. Remove the baking sheet from the oven. Add the pecans and toss. Return to the oven and cook for another 15 minutes.
6. Serve sprinkled with fresh parsley.

Sautéed Arugula with Basil

(My fav, better flavor than spinach!)

Preparation Time: 5 minutes

Cook Time: 10 minutes

Servings: 4

2 tablespoons olive oil
½ teaspoons garlic powder
¼ cup fresh basil
6 cups Arugula
Salt and pepper to taste

1. Heat olive oil in medium heat in skillet. Add garlic and basil and cook for 1 minute.
2. Add arugula to skillet; sauté and toss regularly, until soft and limp.
3. Salt and pepper to taste.

Roasted Japanese Sweet Potatoes

Preparation Time: 5 minutes

Cook Time: 1 hour and 15 minutes

Servings: 6–8 depending on size of potatoes

6 small Japanese sweet potatoes (yams are okay too!)

1 tablespoon ghee

6 sheets of parchment paper

1. Preheat regular oven or toaster oven to 375 degrees F.

2. Scrub and wash potatoes well, and cut off bruised or brown areas (possible moldy areas). The potato skin is very high in potassium, so you will want to eat it, as long as it's clean and free of mold!

3. Take a fork and punch 3-4 times into the potato flesh.

4. Rub thin layer of ghee on potatoes

5. Wrap up each potato in individual parchment paper tightly and twist each end like a candy wrapper. This is to seal up the moistness with the ghee!

6. Roast on baking sheet in oven for 1 hour and 15 minutes. Unwrap and serve.

Serving size is ½ cup.

Breakfast

Eggs Over Easy with "Creamed" Spinach

Preparation Time: 10 minutes

Cook Time: 10–12 minutes

Servings: 2

2 teaspoons coconut oil, plus extra
2 cloves garlic, minced (optional)
2 cups chopped baby spinach
Salt and pepper to taste
2 teaspoons coconut milk
2 eggs

1. In a large skillet, heat the coconut oil over medium low heat.
2. Add the garlic and sauté 1 minute.
3. Turn the heat up to medium high, add the spinach, and season with salt and pepper to taste.
4. Cook the spinach, tossing frequently, until just wilted, 3–5 minutes. Turn off the heat and stir in the coconut milk.
5. Meanwhile cook eggs over easy, with a little bit of coconut oil. Add cooked egg on plate, and spoon on top a dollop of creamed spinach. Sprinkle with more salt, if desired. Serve hot.

Blueberry Lemon Quinoa Porridge

Preparation Time: 5 minutes

Cook time: 25 minutes

Servings: 2

1 cup quinoa

2 cups unsweetened almond milk

1 tablespoon ghee

1 pinch salt

½ lemon peel, grated

1 cup blueberries

2 teaspoons sprouted flax seed

1. Rinse quinoa in a fine strainer with cold water to remove bitterness until water runs clear and is no longer frothy.

2. Heat almond milk in a saucepan over medium heat until warm, 2 to 3 minutes.

3. Stir quinoa, ghee and salt into the milk; simmer over medium-low heat until much of the liquid has been absorbed, about 20 minutes. Remove saucepan from heat. Stir grated lemon peel into the quinoa mixture. Gently fold blueberries into the mixture.

4. Divide quinoa mixture between 2 bowls; top each with 1 teaspoon sprouted flax seed. Serve with unsweetened almond milk on the side.

Pumpkin Pancakes

Preparation Time: 10 minutes

Cook Time: 20 minutes

Servings: 6-8 medium size pancakes

4 eggs, beaten

½ cup pumpkin puree

1–2 tablespoons coconut flour (optional)

1 teaspoon pure vanilla extract

6 drops alcohol-free liquid stevia

1 teaspoon pumpkin pie spice (½ teaspoon cinnamon, ¼ teaspoon ground ginger, ⅛ teaspoon nutmeg)

1 pinch Celtic salt

¼ teaspoon baking soda

2 tablespoons ghee or coconut oil

Coconut or avocado oil for cooking

Cinnamon powder

1. Whisk the eggs, pumpkin puree, coconut flour (optional), pure vanilla extract, and liquid stevia. Sift the pumpkin pie spice, salt and baking soda into the wet ingredients.

2. Melt ghee or coconut oil in a large skillet over medium heat and mix into batter.

3. In the same skillet, add 1 tablespoon of avocado or coconut oil.

4. Spoon the batter into the skillet to make medium sized pancakes, about ¼ cup. When a few bubbles appear, flip the pancakes

once to finish cooking. Repeat with remaining batter, adding a little oil to the skillet before cooking each pancake.

5. Serve with ghee and cinnamon powder on top!

Zucchini Pancakes

Preparation Time: 10 minutes

Cook Time: 15 minutes

Yield: 10 to 12 pancakes

4 large eggs
3 cups grated zucchini
¼ cup chopped green onions
¾ to 1 cup almond flour
½ teaspoon Celtic sea salt
Freshly ground black pepper
Avocado oil or ghee for cooking

1. Mix all ingredients except the oil together in a medium-sized bowl.

2. Heat a 10-inch stainless steel skillet or cast iron skillet over medium-low heat. Be sure to heat your pan long enough before adding the oil and batter; otherwise the pancakes will stick. Add about 1 tablespoon avocado oil or ghee.

3. Spoon the batter into hot skillet, about ¼ cup. Cook for a few minutes on each side. When a few bubbles appear, flip the pancakes once to finish cooking. Repeat with remaining batter, adding a little oil or butter to the skillet before cooking each pancake.

Goes well with turkey bacon or sausages.

Veggie Scramble

Preparation Time: 5 minutes

Cook Time: 5 minutes

Servings: 6

Avocado or coconut oil for sautéing

½ cup sliced red bell pepper

¼ cup chives

2 cups arugula

6 large eggs

2 tablespoons cilantro, chopped

¼ avocado, sliced, per serving

1. Heat oil in skillet over medium heat. Add red peppers, chives and arugula and sauté until tender.
2. In a medium mixing bowl, whisk eggs and cilantro. Pour over peppers and arugula and cook until done.
3. Garnish with sliced avocado.

Soup and Salad

Dr. Susanne's TOC (Tomato Olive Cucumber) Salad

Preparation Time: 10 minutes

Servings: 4

1 pint cherry tomatoes, halved (**Note**: I cut off the end of each tomato, this is where it can get moldy)

½ cup Kalamata olives, halved

1 large English cucumber, chopped into 1 inch pieces

3–4 tablespoons extra virgin olive oil

½ lemon, juiced

½ teaspoon dried oregano

Sea salt and black pepper to taste

Toss all ingredients together in a small mixing bowl. Serve room temperature or chilled.

Arugula-Endive-Radicchio Salad

Preparation Time: 10 minutes

Servings: 4

Salad Ingredients:

5 cups arugula

½ radicchio chopped

1 endive chopped

1 tablespoon chives

1 carrot grated

¼ cup red bell pepper, chopped

White Balsamic Basil Vinaigrette Dressing Ingredients

⅓ cup extra virgin olive oil

¼ cup white balsamic vinegar

6 fresh basil leaves

½ teaspoon dried oregano

¼ teaspoon Celtic sea salt

¼ teaspoon black pepper

¼ teaspoon red chili flakes

1. Toss all salad ingredients in a large salad bowl
2. Add all dressing ingredients into a small cup made for a personal blender, and use the 4-prong blade.
3. Blend all ingredients at high speed until emulsified (thick liquid)
4. Drizzle dressing over salad and serve immediately

Vegan Italian Soup

Preparation Time: 10 minutes

Cook Time: 45 minutes – 1 hour

Servings: 6

Cooking accessory: spice bag or muslin bag with drawstring for onion and garlic

4 tablespoons avocado or olive oil, divided
1 cup diced carrots
1 cup diced yellow squash
1 cup diced celery
1 cup chayote (optional)
1 tablespoon dried oregano
½ teaspoon dried or fresh rosemary
2 tablespoon chopped basil
2 tablespoon Italian parsley
6 cups low-sodium vegetable broth
1 medium white or yellow onion, cut in quarters
2 cloves garlic halved insert to muslin bag (flavors without increasing fermentable carbohydrates to the soup)
Celtic sea salt and pepper to taste
Fresh chopped basil to garnish

1. Heat 2 tablespoons avocado or olive oil in large pot over medium-high heat. Add carrot, squash, celery and chayote. Cook 5 minutes more.

2. Add oregano, rosemary, basil and parsley. Cook 1 minute. Stir in vegetable broth. Add garlic and onion to spice or muslin bag, add to pot and bring to a boil.

3. Simmer until vegetables are thoroughly cooked. Salt and pepper to taste. Pull out bagged onion and garlic.

4. Garnish with basil before serving.

Roasted Tomato Basil Soup

Preparation Time: 15 minutes

Cook Time: 1 hour, 35 minutes

Servings: 6–8

Cooking accessory: 1) spice bag or muslin bag with drawstring for onion and garlic 2) immersion blender

3	pounds ripe plum tomatoes, cut in half lengthwise
¼	cup and 2 tablespoons olive oil, divided
1	tablespoon Celtic sea salt
1½	teaspoons freshly ground black pepper
2	tablespoons ghee butter
¼	teaspoon crushed red pepper flakes
1	28-ounce canned plum tomatoes, with juice
4	cups fresh basil leaves, packed
1	teaspoon fresh thyme leaves
1	teaspoon dried oregano
1	yellow onion, cut in quarters
6	garlic cloves, halved
1	quart vegetable or chicken stock or purified water

1. Preheat the oven to 400 degrees F. Toss together the tomatoes, ¼ cup olive oil, salt, and pepper. Spread the tomatoes in 1 layer on a baking sheet and roast for 45 minutes.

2. In an 8-quart stockpot over medium heat warm 2 tablespoons of olive oil, ghee, and red pepper flakes for 1 minute.

3. Add the canned tomatoes, basil, thyme, and oregano. Add onion and garlic to a spice bag or muslin bag and place in stockpot along with liquid vegetable stock.

4. Add the oven-roasted tomatoes with its juices and bring to a boil and simmer uncovered for 40 minutes.

5. Use immersion blender and blend into a tomato soup puree. Taste for seasonings. Serve hot or cold.

Ox Tail Bone Broth Soup

Preparation Time: 30 minutes

Cook Time: 8-24 hours

Servings: 8

Cooking accessory: spice bag or muslin bag with drawstring for onion and garlic

- 4 pounds ox tail or other mixture of marrowbones and bones with a little meat on them, such as short ribs, or knuckle bones (cut in half by a butcher)
- 2 unpeeled carrots, cut into 2-inch pieces
- 1 whole leek, cut into 2-inch pieces
- 12 cups purified water
- 2 celery stalks, cut into 2-inch pieces
- 2 bay leaves
- 2 tablespoons black peppercorns
- 1 tablespoon cider vinegar
- 1 medium onion, quartered
- 1 garlic head, halved crosswise

1. Preheat oven to 450 degrees F. Place beef bones, carrots and leek on a parchment paper lined roasting pan or rimmed baking sheet and roast for 20 minutes. Toss ingredients and continue to roast until deeply browned, 10 to 20 minutes more.

2. Fill a large (at least 6-quart) stockpot with 12 cups of purified water. Add celery, bay leaves, peppercorns, and vinegar and bagged onion and garlic.

3. Scrape the roasted bones and vegetables into the pot along

with any juices. Add more water if necessary to cover bones and vegetables.

4. Cover the pot and bring to a gentle boil. Reduce heat to a very low simmer, skimming foam and excess fat occasionally, for at least 8 hours. The longer you simmer, the better the stock. You can simmer up to 24 hours.

5. Remove the pot from the heat and let cool slightly. Strain broth using a fine-mesh sieve and remove bones and vegetables (you can eat the leftover meat off the oxtail with the vegetables and the broth for lunch!). Continue to let the broth cool to room temperature and then refrigerate in smaller glass containers overnight. Remove solidified fat from the top of the chilled broth before serving. You can freeze for up to 6 months.

with any juices. Add more water if necessary [illegible] bones and vegetables.

[illegible] over the pot and simmer [illegible] very low simmer [illegible] at least [illegible] hours [illegible]
[illegible] simmering [illegible]

4. [illegible] the heat [illegible] using a fine mesh sieve and [illegible] bones [illegible] and the [illegible] [illegible]

Slow Cooking

Crock Pot Paleo Osso Bucco

Preparation Time: 15 minutes

Cook Time: 8 hours

Servings: 4

Cooking accessory: spice bag or muslin bag with drawstring for onion and garlic

2 lbs. lamb or veal shanks (osso buco)
1 32-ounce jar organic marinara sauce (includes seasonings)
1 yellow or white onion, quartered
4 garlic cloves, halved
2 large carrots, peeled and roughly chopped
3 tablespoons chopped basil
½ teaspoon dried rosemary
Celtic sea salt and **black pepper** to taste

1. Place frozen or fresh veal or lamb, marinara sauce, carrots, bagged onion and garlic, basil, rosemary, salt, and pepper into the slow cooker **on low setting for 8 hours** (overnight or all day). The dish is done when the meat is falling off of the bone.

2. Serve with Sautéed Arugula

Slow Cooker Beef Stew

Preparation Time: 20 minutes

Cook Time: 8 hours

Servings: 6

Cooking accessory: spice bag or muslin bag with drawstring for onion and garlic

1½ pounds grass-fed beef chuck
½ teaspoon sea salt
½ teaspoon freshly ground black pepper
3 tablespoons organic tomato paste
1 tablespoon balsamic vinegar
1⅓ cups beef stock
1 bay leaf
¼ teaspoon dried rosemary
1 medium yellow onion, quartered
3 cloves garlic, halved
1 peeled chayote or zucchini, cut into 1-inch pieces
4 medium carrots, cut into 1-inch pieces
½ pound green beans, trimmed and cut into 2-inch lengths
1-2 tablespoon arrowroot powder, to thicken (optional)

1. Cut beef chuck into 1½-inch chunks. Place the beef chunks in a 6-quart slow cooker and generously season with sea salt and black pepper.

2. Add the tomato paste, vinegar and beef stock.

3. Add bay leaf and rosemary. Add onion and garlic to spice or muslin bag and add to slow cooker.

4. Add chayote or zucchini and carrots. Do not stir.

5. Cover slow cooker; cook on high until beef is fork-tender, about 5 hours or cook on low heat for about 7–8 hours.

6. During last 30–45 minutes of cook time, stir the green beans into the stew. If you'd like to thicken up the broth, just before serving, add about 1–2 tablespoons of arrowroot powder and stir thoroughly.

7. Serve over a half cup of quinoa or brown rice.

Slow Cooked Bean-Free Chili

Preparation Time: 20 minutes

Cook Time: 8 hours

Servings: 4–6

Cooking accessory: spice bag or muslin bag with drawstring for onion and garlic

Seasoning Mix:

4 tablespoons chili powder

2½ teaspoons ground coriander

2½ teaspoons ground cumin

1½ teaspoons garlic powder

1 teaspoon oregano

½ teaspoon cayenne pepper

Chili:

1½ pounds ground turkey or beef

1 peeled chayote or zucchini, cut into 1-inch pieces

3–4 cups arugula

1 small yellow onion, quartered

3 cloves garlic halved

1 28-ounce can diced tomatoes

1 15-ounce can tomato sauce

1. Mix together seasoning mix. You will only need 5 teaspoons for this recipe; save the rest for another time. Store the extra seasoning in an airtight container in a cool, dry place.

2. In a skillet, cook ground meat thoroughly. Drain.

3. Add 3 teaspoons of seasoning mix to ground meat.

4. In the slow cooker, add arugula, tomatoes, tomato sauce, chayote or zucchini, bagged onion and garlic, and 2 more teaspoons of seasoning mix.

5. Add prepared meat to the slow cooker.

6. Stir together and cook on low setting for 6 to 8 hours.

7. Great over quinoa or on top of a small sweet potato.

Cilantro Curry Chicken

Preparation Time: 20 minutes

Cook Time: 8 hours

Servings: 4–6

Cooking accessory: spice bag or muslin bag with drawstring for onion and garlic

11/2 lb. chicken- legs, thighs, breast, cut up as desired
1 cup carrots, cut into 1/2-inch pieces
1 cup chopped celery
1 small yellow onion, quartered
5 garlic cloves, halved
1 cup chopped cilantro
1 cup diced Japanese sweet potato or yam
16 ounces chicken broth
2 teaspoons yellow curry spice
Celtic salt to taste
Ghee for coating the bottom of the ceramic slow cooker
1 tablespoon arrowroot (optional, used to thicken sauce)

1. Lightly coat the bottom of the slow cooker with ghee or healthy oil such as avocado or coconut oil.

2. First add the chicken, then carrots, sweet potato and celery on top of the chicken and the bagged onion and garlic. Pour chicken broth on top until you cover the ingredients.

3. Stir in curry and cilantro.

4. Cover and turn on to low setting for 8 hours or overnight.

5. Optional: If you'd like to thicken up the sauce, just before serving, add about 1–2 tablespoons of arrowroot powder and stir thoroughly.

6. Perfect over sticky sweet brown rice or quinoa.

Bonus Dessert Recipe

Mint and Strawberry Cobbler

Preparation time: 20 minutes

Cook time: 40 minutes

Servings: 4

Special equipment:

10″ cast iron skillet or medium-sized baking dish (8×8″ or 9×9″).

Filling ingredients:

1 pound strawberries, fresh or thawed

2 tablespoons lime juice

Grated peel of one lime

1 tablespoon mint, finely chopped from about 20 leaves

1 tablespoon ginger, grated

1 teaspoon vanilla powder or extract

2 teaspoons arrowroot or tapioca starch

Celtic sea salt, to taste

Topping ingredients:

1 cup finely shredded coconut flakes, unsweetened

½ cup ghee or coconut oil, melted

¼ cup arrowroot or tapioca starch

½ teaspoon vanilla powder

Celtic sea salt, to taste

1. Preheat the oven to 350 degrees F.

2. Combine all the filling ingredients and toss to evenly cover the strawberries. Place in the cast iron skillet or baking dish.

3. Combine all the topping ingredients in a separate bowl and use your hands to work the ghee/coconut oil well into the paste until it turns into a crumble.

4. Spread the crumble evenly on top of the strawberries.

5. Bake for 30 to 40 minutes or until the topping is brown

Plan Summary

What steps do you need to get started? Here's a summary for you that you can make reference to, as you work to put the Mighty Mito Plan in action.

Change Your Environment

Look at your environment and see what you can do to change things in order to help protect your mitochondria and prevent future damage.

First, you'll want to adjust some of your habits.

- Quit smoking
- Stay in your car for a few minutes before you get out, especially when pulling into a garage or underground parking
- Leave the garage door open long enough to release carbon monoxide after pulling your car in
- Stop drinking alcohol (or limit to one drink a week, and only specific types of alcohol with no gluten or sugar)
- Avoid antibiotics when they're not essential for recovery and avoid statin drugs
- Stop using plastic containers to store and heat food in

Next, here are some things that you can do to clean up your environment and in turn, clean up your mitochondria:

- Get a HEPA filter for each room in your home
- Add air-cleansing plants around your house
- Throw out your old makeup and beauty products
- Buy new beauty products that are made from all natural ingredients
- Get rid of air fresheners, perfume, potpourri, and other VOCs
- Buy organic produce whenever available
- Buy a reverse-osmosis water purifier for your home or tap

Burst to Boost™ training

Once you've changed your environment, the next step is to change your health through a mitochondria-healthy diet and mitochondria-approved exercise.

When you exercise, you want to do aerobic exercise that brings oxygen to your muscles and the many mitochondria found there. This is done with the Burst to Boost™ training.

- Choose 3 days a week to exercise; nail down a time of day when you have 20 to 30 minutes to exercise. These days need to be spread out with a rest day in between (e.g., Sunday, Tuesday, Thursday; Monday, Wednesday, Friday; Tuesday, Thursday, Saturday)

- Start at your current fitness level, whatever that is, and choose an activity that fits it
- For the activity you choose, burst for one minute (if you're able). Take a rest period (for as long as you need to, but one minute is optimal) and then burst again. The goal is to do this for 20 minutes (again, if you're able) and 30 minutes at most.

Ideas that work well for Burst to Boost™ training:

- Treadmill
- Walking/jogging/running
- Stationary bike
- CrossFit
- Climbing stairs
- Jump roping
- Weight training
- Hula hooping
- Riding a bike
- Swimming
- Lunges and squats (walking in between bursts)
- Sit-ups and pushups (stretching and walking in between bursts)

Here's what you need to get started:

- Equipment for your choice of exercise
- Time keeper, this could be a stopwatch, watch, or timer. You want to ensure you don't go past 60 seconds (though listen to your body. You should push yourself to about 85% of your max)
- Comfortable clothes that won't be too hot
- Supportive shoes appropriate for the activity

And don't forget that you can do exercise that involves stretching and slower movements on your off days, such as tai chi, restorative yoga, qi gong, and the like to help with flexibility and muscle tone.

Mighty Mito Nutrition Plan

Along with exercise, it's time to overhaul what you eat. This isn't a diet, it's a new way of looking at your nutrition. If you're serious about producing healthy mitochondria, this isn't a short-term fix but a lifestyle that will benefit you throughout your life. But the best part about this plan is if you follow it, you won't have to worry about counting calories or limiting portions.

Food Elimination

Start by eliminating the following foods:

- Foods with GMO ingredients
- Non-organic foods
- Rancid and denatured oils

- Fried foods
- Trans fats and partially hydrogenated fats
- Sulfite drenched dried fruits (apricots, raisins, mangos, etc.)
- Fungal contaminated foods (cheese, dried fruits, peanuts, cashews and pistachios, packaged snacks, bruised and overripe fruits, old veggies)
- Food coloring and food additives
- Artificial sweeteners
- Dairy from animal sources in all forms (milk, yogurt, ice cream, cheese)
- Gluten (in wheat, barley, rye, kamut, spelt)
- Sugar and processed sweeteners (white sugar, brown sugar, agave, and honey)
- Processed foods (if you're not sure, take a look at the ingredients. If there is an ingredient you can't pronounce, it's processed).

I recommend that you remove the above foods completely from your diet because of the effects they have on the human body. The following is a list of food you will remove from your diet for a short period of time to cleanse the body. You will then reintroduce them slowly (adding one every 4 days) and keep track of how you feel with each introduction. This will tell you whether your body does well with that food or not.

- Beans and legumes (including peanuts, cashews, and soy)
- Vegetables high in fermentable carbohydrates

- Fruit high in fermentable carbohydrates
- Avocados
- Apples
- Broccoli
- Asparagus
- Sweet green bell peppers
- Mushrooms
- Cabbage
- Onions
- Brussels sprouts

This is not a comprehensive list see the Ultimate Wellness Food Checklist for a complete list of what you need to avoid for the first 6 weeks and what you can eat (which is a long list!).

If you need a reminder of why I'm having you remove these foods, see "Why Can't I Eat these foods?" in chapter 11.

Focus on healthful foods including the following:

- Free-range beef, pork, chicken, and lamb
- Free-range eggs
- Fish (1 time a week to limit mercury and nuclear waste intake)
- Zucchini
- Arugula
- Amaranth
- Spinach

- Bok choy
- Swiss chard, rainbow chard, and kale
- Tomatoes
- Blueberries
- Red bell peppers
- Quinoa
- Sweet potatoes
- Almond, coconut, and rice milk
- Walnuts, almonds
- Pumpkin seeds

Again, this is not a comprehensive list. The food you eat should be organic and all meat should be free-range or wild caught. See the list of recipes for ideas on how to explore these new foods.

Healthful Supplements

Last, get the following supplements and add them to your daily intake. You may be able to get a combination of some of these in a mitochondrial formula. Whatever you choose, make sure they are quality products that are made from organic ingredients:

- Multi-vitamin/mineral
- Glutathione (GSH)
- L-carnitine
- CoQ10 Enzyme
- Essential Amino Acids

- Mitochondrial Membrane Lipid Therapy: Phosphatidylcholine and Omega 3 Fatty Acids (Docosahexaenoic acid, DHA)
- Magnesium
- Pyrroloquinoline quinone (PQQ)
- Buffered Creatine Monohydrate
- Rhodiola rosea
- Alpha-Lipoic Acid (ALA)
- NADH: Reduced Nicotinamide Adenine Dinucleotide
- Astaxanthin
- Vitamin D
- Vitamin C
- Vitamin E
- NAC (N-Acetyl Cysteine)
- Melatonin (take about 30 minutes before bedtime)
- D-Ribose
- Shilajit

Dairy-Free/Paleo Oxygenating Pudding

In addition to improving your diet and adding supplements, the next thing you'll want to do to improve your mitochondria is eating oxygenated pudding. This is easily absorbed by your body and will increase your energy and sharpen your brain. I recommend a dairy-free version. Here's the recipe:

- 4 tablespoons unsweetened almond or coconut milk or unsweetened non-dairy coconut yogurt
- One scoop egg white protein, either vanilla or chocolate (hormone-free, antibiotic-free with no artificial flavors or sweeteners. Can be sweetened with stevia)
- 2 tablespoons high lignan flax oil (I recommend using Bar leans brand)

Add the above ingredients together in order to a glass cup or mug narrow enough to submerge an immersion blender. Blend together with an immersion blender on low setting pulsing on and off slowly to prevent heating the mixture. Move blender up and down as you pulse. Continue for about one minute till you can't see any oil droplets.

Variations: Add blueberries, raspberries, or freshly ground flaxseed powder.

Emotional and Psychological Health

Along with physical and nutritional health, there are new habits you can do every day to improve your emotional and mental health as well. Here's your checklist:

- Practice belly breathing three times a day for 10 minutes each time
- Practice the laughter exercise twice a day
- An hour before bed, dim the lights, turn off all electronics, and read a book (not on a tablet) using a small book light
- Don't drink alcohol or caffeine 6 to 7 hours before bed time

- Don't get in bed till you're ready to go to sleep
- If you suffer from insomnia and don't fall asleep after 20 minutes, get up and do something and then return to bed when you feel sleepy

Looking for more resources and help?

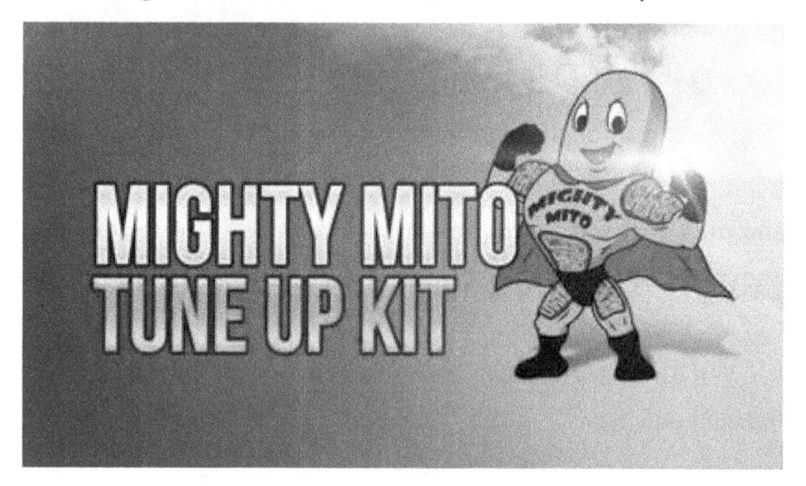

There are only so many things that can be included in a book. And many things are much easier to be *shown* than read in a book.

So I've compiled a few audio and video resources that will help you take your health to the next level even quicker. The Mighty Mito Tune Up Kit is a great program to complement this book. For more details, go to the Resources section.

Resources

Private Practice

Dr. Susanne Bennett

Wellness for Life Center

1821 Wilshire Blvd., Suite 300

Santa Monica, CA 90403

Tel: 310-315-1514

Fax: 310-315-1504

email: dr@drsusanne.com

Dr. Susanne's Websites

drsusanne.com

purigenex.com

wellnessforliferadio.com

the7dayallergymakeover.com

mightymito.com

Dr. Susanne's Social Media

Facebook: facebook.com/drsusannebennettallergyspecialist

Instagram: @drsusanne

LinkedIn: Dr. Susanne Bennett

Pinterest: @drsusanne

Twitter: @drsusanne

Links & Downloads

1. The 7-Day Allergy Makeover by Dr. Susanne Bennett http://the7dayallergymakeoverbook.com
2. Mitochondrial Dysfunction and Molecular Pathways of Disease by Dr. Steve Pieczenik and Dr. John Neustadt http://www.nbihealth.com/publications/mitochondrial_dysfunction.pdf
3. Ultimate Wellness Food Checklist (UWFC): http://drsusanne.com/uwfc
4. Mighty Mito Tune Up Kit: http://drsusanne.com/tuneup
5. Burst to Boost™ training: http://drsusanne.com/tuneup
6. Sacred Animal Qigong: http://drsusanne.com/tuneup
7. Essential Nutrients: Quick Reference Guide: http://drsusanne.com/mmnutrients
8. Power of Glutathione audio: http://drsusanne.com/GSH
9. Super 8 Aminos: http://drsusanne.com/super8
10. *Lipid Replacement and Antioxidant Nutritional Therapy for Restoring Mitochondrial Function and Reducing Fatigue in Chronic Fatigue Syndrome and other Fatiguing Illnesses* by Dr. Garth Nicholson and Dr. Rita Ellithorpe: http://www.immed.org/publications/Nicolson_ElllithorpeJCFS_copy.pdf
11. Vital Choice Seafood: http://drsusanne.com/vitalchoice
12. Jing Jing powder with Shilajit: http://drsusanne.com/product/jing-jing/
13. *Laughing for Life* exercise video: http://drsusanne.com/laugh
14. Wellness for Life Supplements: http://drsusanne.com/store/

References

[1] Stokel, Kirk. "Rejuvenate Your Cells by Growing New Mitochondria." http://www.lifeextension.com/magazine/2010/ss/rejuvenate-your-cells-growing-new-mitochondria/page-01. Accessed 12-4-2015

[2] Beil, Laura. "Medicine's New Epicenter?" http://www.curetoday.com/publications/cure/2008/winter2008/medicines-new-epicenter-epigenetics?p=2 Accessed 12/1/2015.

[3] Nicolson G. Altern Ther Health Med. 2014; 20 (suppl 1): 18-25

[4] Quillin, Patrick. "Cancer's Sweet Tooth" http://www.mercola.com/article/sugar/sugar_cancer.htm. Retrieved 12/4/15.

[5] Byun HM, et al. Part Fibre Toxicol. 2013 May 8;10:18.

[6] Kalghatgi S, et al. Sci Transl Med. 2013 Jul 3;5(192):192ra85

[7] Lin Y, et al. Cell Death Dis. 2013 Jan 17;4:e460.

[8] Kreher JB, Schwartz JB. Overtraining Syndrome: A Practical Guide. *Sports Health*. 2012;4(2):128-138. doi:10.1177/1941738111434406.

[9] Friden J, Seger J, Ekblom B. Sublethal muscle fiber injuries after high tension anaerobic exercise. Eur J Appl Phys 1988; 57:360-368.

[10] Strickertsson, JA, Desler, C, Rasmussen, LJ. (2014 August). Impact of bacterial infections on aging and cancer: impairment of DNA repair and mitochondrial function of host cells. Retrieved from http://www.ncbi.nlm.nih.gov/pubmed/24704713.

[11] http://antimicrobe.org/m02.asp

[12] http://antimicrobe.org/m02.asp

[13]Abidov, M.; Crendal F.; Grachev, S.; Seifulla, R.; Ziegenfuss, T.; (2003 December). Effect of Extracts from Rhodiola Rosea and Rhodiola Crenulata (Crassulaceae) Roots on ATP Content in Mitochondria of Skeletal Muscles. Retrieved from http://link.springer.com/article/10.1023%2FB%3ABEBM.0000020211.24779.15

[14] Forsyth, LM. et al. (1999 December). Oral NADH for Chronic Fatigue Syndrome. Retrieved from http://www.ahcmedia.com/articles/50590-oral-nadh-for-chronic-fatigue-syndrome.

[15] Srinivasan V et al. Int J Alzheimers Dis. 2011; 2011: 326320. Published online 2011 May 4. doi: 10.4061/2011/326320

Printed in the USA
CPSIA information can be obtained
at www.ICGtesting.com
LVHW020839111123
763663LV00043B/1501